HYDROELECTRICITY, ENVIRONMENT AND QUALITY OF LIFE

HYDROELECTRICITY, ENVIRONMENT AND QUALITY OF LIFE

ASHOK KUMAR TIWARI

REGAL PUBLICATIONS
New Delhi - 110 027

HYDROELECTRICITY, ENVIRONMENT AND QUALITY OF LIFE

ISBN 978-81-8484-050-6

Typeset by
RAHUL COMPOSERS
358, Pocket-B, Phase-II, Sector-16 B, Dwarka, New Delhi - 110 075

Printed in India at
MAYUR ENTERPRISES
WZ Plot No. 3, Gujjar Market, Tihar Village, New Delhi - 110 018

Published by
REGAL PUBLICATIONS
F-159, Rajouri Garden, New Delhi - 110 027 • Phone : 45546396
E-mail : regalbookspub@yahoo.com

Dedicated to

The Memory of

My Mother

Contents

Appendices :

Preface

The most important single factor which can act as a constraint on economic growth is the availability of energy. It is universally accepted that energy is a vital infrastructure for economic development. Hydroelectricity is by far the largest renewable resource used for electricity generation. India has completed 58 years with development works in each sphere including industry, agriculture and power development. Now that the country is moving with a faster pace of industrial development with globalization, it has to ensure availability of reliable power. Though India has made rapid strides in capacity addition programme after independence from mere 1364 MW in 1947 to about 1,10,000 MW in 2005, the country is still suffering from power shortages and load shedding. This is hampering the economic growth in industrial, agricultural and other sectors. There is an urgent need to increase power generation for industrial development as well as to improve the quality of life of the vast population. It is in this background that the present study was set to study the relationship between hydroelectricity and economic development.

The description of the study has been presented in six chapters. *Chapter 1* highlights the hydropower scenarios and research design of the study. *Chapter 2* emphasizes the role of hydroelectricity in the process of economic development, quality of life and rural development. In this chapter plan efforts for the development of power sector in India as well as in Himachal Pradesh have also been discussed. *Chapter 3* deals with the study of hydropower projects on the Satluj River Valley in respect of implementation of the resettlement action plan, socio-economic status of project affected families, etc. This chapter also examines the role of Hydropower projects of Satluj Jal Vidyut Nigam in improving quality of life in the vicinity of the hydropower projects in particular and in the region in general. *Chapter 4* has been devoted to study the environment concerns of hydropower projects. It includes utilization of available natural resources for development, criticism of large hydro-projects on environmental considerations, problems of displacement and rehabilitation of project affected families and benefits of hydropower resources. *Chapter 5* emphasizes the socio-economic development in the Northwestern Himalayan region (Himachal Pradesh). Summary, conclusion and policy implications of the study have been given in *Chapter 6*.

The author expresses a deep sense of gratitude to the revered scholars, researchers and experts who have imparted very useful advice, constructive suggestions and valuable support at various stages of the study, especially, Er. Deepak Nakhashi, Deputy General Manager, SJVNL, Shimla; Professor H.S. Parmar, Department of Economics, ICDEOL; Professor Yoginder Verma, Director, UGC-Academic Staff College, Professor B.S. Marh (Former Director, IIHS), Department of Geography, Dr. Shyam Prasad Sharma, Chairman, Department of Economics; Prof. N.S. Bist, Department of Economics, Dr. P.D. Bhardwaj, Department of Geography, Dr. K.R. Pant, Librarian (all from Himachal Pradesh University), Dr. Krishna Mohan, Chairman, Department of Geography, Punjab University, Chandigarh. I am greatly indebted to learned authors, scholars whose work I have consulted and quoted for giving the book its present shape. I am thankful to my family members for their constant encouragement and support.

My thanks are due to Shri R.D.S. Bhatia, M/s. Regal Publications, New Delhi for taking all pains for bringing out this book.

ASHOK KUMAR TIWARI

1

Introduction

This chapter has five important objectives. First, it gives world's hydroelectric scenario. Secondly, it highlights India's hydroelectric scenario. Thirdly, it deals with governmental initiatives towards accelerated hydropower development in India. Fourthly, it discusses hydropower scenario in the western Himalayan region. The last part of this chapter provides the data base of the study as well as research methodology used for the measurement of degree of relationship between power sector and economic development including the measurement of development index.

Hydroelectric power commonly known as electricity generated using the potential energy of water. It is a renewable energy source with considerable potential worldwide, even though accounts for only a small proportion of the world's energy requirements. Renewables are at times described as the *"dreams of the 1970s, realities but luxuries of 2000, and the necessities of 2020 and thereafter"*. Renewables include Hydropower, Biomass, Solar, Wind, Geothermal and Ocean Sources. Hydropower has certain advantages, most important among them being the ability to start and stop quickly and instantaneous load acceptance or rejection. This makes it particularly suitable to meet peak demand and for enhancing system reliability and stability. Some of the other advantages of hydropower are long life of the hydropower plants, renewable character of the energy source, very little operating and maintenance costs, absence of inflationary pressures experienced by the fossil fuels, etc. Like most renewable sources, hydroelectricity plants are highly capital intensive, but have lower operational costs than the thermal or the nuclear option. The high opening cost has been the hurdle for its growth in most of the developing countries like India, where bulk of the unexploited potential is located. Hydropower has certain disadvantages also, though it is considered as a clean energy source, it is not absolutely devoid of greenhouse gas emissions, ecosystem burdens and socio-economic adversities.

HYDROELECTRICITY

1.1 WORLD'S HYDROELECTRIC SCENARIO

Electricity is most helpful gizmo of modern society[1] and is sometime called as man's most constructive servant. Hydroelectricity is undoubtedly the largest renewal resources used for electricity generation. The worldwide theoretical, technical and economic potential of hydropower have been estimated[2] as theoretical potential at about 40,500 Tera Watt Hours, technical potential at about 14,320 TWh, and economic potential at about 8100 TWh. Currently, about one-third of the economic potential has so far been developed. The US and Western Europe have developed about 76 per cent and 65 per cent of their potential respectively. Whereas countries like Africa and Asia have only developed about 20 per cent of their potential. The world installed hydro-capacity stands at 694,000 MW. Canada, Brazil, US, China and Russia were the top five producers of hydroelectric power and their collective hydro-power generation accounted for 49 per cent of the world total. Worldwide about 125,000 MW of hydro-capacity is under construction, bulk of which is in the developing countries. China, India, Malaysia and other developing Asian countries have striving plans of hydropower capacity additions.

The word's Total Primary Energy Consumption in 2000 was 9096 million tones of oil equivalent, and with a world population of 6056 million, it translates into global average per capita consumption of about 1502 kg. The regional variations are large. North American average is at 6547 kg. of oil equivalent while in African region it is just 349 kg. of oil equivalent per capita. Fossil fuels account for over 85 per cent of the total energy consumed. The trend in Energy Mix in the world is given in Table 1.1.

TABLE 1.1

Trends in Energy Mix (%) in the World (1970-2000)

Year	*Oil*	*Coal*	*Natural Gas*	*Hydro/ Other Renewables*	*Nuclear*	*Total Energy*
1970	45%	31%	18%	5%	0%	100%
1980	45%	27%	20%	6%	2%	100%
1990	39%	28%	22%	6%	6%	100%
2000	39%	24%	24%	7%	6%	100%

Source : Compiled from Reliance Review of Energy Markets, 2002, p. 14.

Table 1.1 reveals that initial during 1970, the share of oil energy was highest at 45 percent and nuclear energy was almost negligible (i.e. zero per cent). But, during the year 2000, the share of nuclear energy rose to 6 per cent whereas hydro and other Renewables energy source increased marginally during 1970-2000.

In the matter of Installed capacity of hydroelectricity in various countries of the world, it is evident from Table 1.2 that the countries like Norway (99%), Brazil (86%), Switzerland (72%), Venezuela (62%) and Canada (60%) are exploiting the maximum share of hydroelectric potential. Most of the countries are now realizing that the fossil fuel sources will be exhausted, and only Renewables sources would be the substitute, particularly the hydroelectricity. The longevity of the hydropower plants, the renewable nature of the energy source, very low operating and maintenance costs, absence of

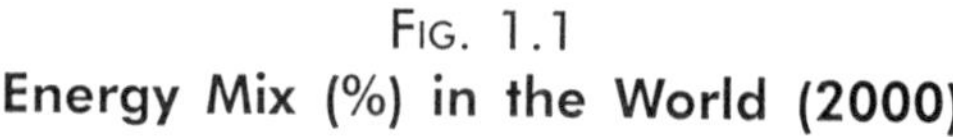

FIG. 1.1
Energy Mix (%) in the World (2000)

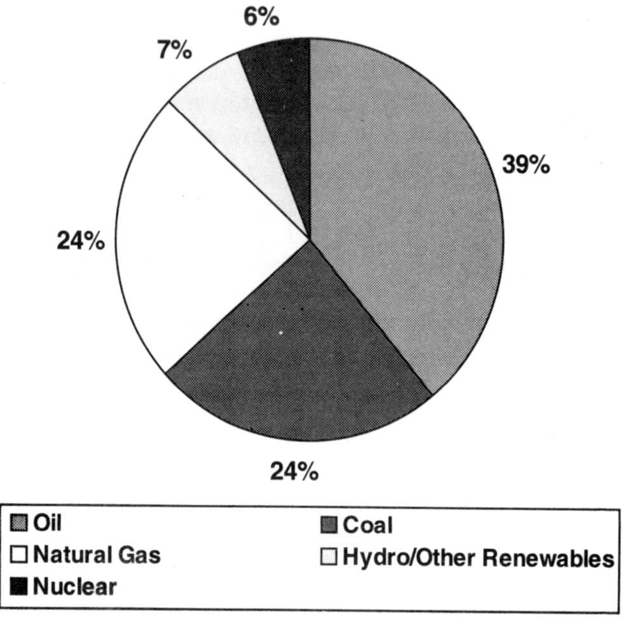

inflationary pressures experienced by the fossil fuels are some of the advantages of hydropower.[3]

Other renewable sources like Biomass, Wind energy, solar energy and geothermal energy, if exploited could also fill the gap of demand and supply of energy in most of the developed and developing nations. A brief sketch of these Renewables is given below:

> Biomass is the term used to describe all plant derived material. It may be used to generate energy by direct combustion or by conversion to either a liquid or a gaseous fuel. Plant materials use the sun's energy to convert atmospheric carbon dioxide to sugars during photosynthesis. On combustion of the biomass, energy is released as the sugars are converted back to carbon dioxide. Thus, energy is harnessed and released in a short

time frame, making biomass energy a renewable energy source. Fossil fuels have also ultimately been derived from atmospheric carbon dioxide, as they are degraded residues of plant and animals sources. However, the time frame is very long of the order of millions of years rather than a few years, as in the case of biomass. The world derives a little over 10 per cent of its energy from Biomass.[4] In the developing and the poorest countries, Biomass is the most important energy source. In the two largest developing countries, China and India, Biomass accounts for almost 20 per cent and 40 per cent respectively of the primary energy supply. In some of the world's poorest countries, Biomass accounts for up to 90 per cent of the energy supply, mostly in traditional/non-commercial forms. The resource potential of Biomass energy is much larger than the current world energy consumption. But given the low conversion efficiency of solar to biomass energy (i.e. less than one per cent), large areas are needed to produce modern energy carriers in substantial amounts.

Wind Energy was extensively used as a source of power before the industrial revolution. It was displaced by the fossil fuels because of costs and reliability. The oil shocks of the 1970s again saw renewed interest in wind energy[5] for applications like grid-connected electricity, water pumping and power supply in remote areas. The United States possessed about 95 per cent of the world's installed capacity of wind power. The installed capacity[6] was estimated to be about 15,000 MW, with 29 countries having active wind energy programs. The estimated electricity generation from wind turbines was about 24 TWh in 1999.

The earth continuously receives a power input of 1.73×10^{14} KW from the sun. This translates to 1.5×10^{18} KWh/year, which is about 10,000 times the world's current annual energy consumption. Therefore, the solar energy has vast theoretical potential. Several factors like daily, seasonal and geographical variations; weather conditions and land availability make the possible potential of solar energy[7] much lower. A UNDP assessment shows that even one per cent of the world's desert area used by solar thermal power plants would be sufficient to

TABLE 1.2
Hydroelectricity Installed Capacity in Various Countries of the World

(As on January 1, 2000) *(in MW)*

Country	*Hydroelectricity*	*Total Energy*	*% share of Hydroelectricity*
(1)	*(2)*	*(3)*	*(4)*
US	99,000	795,000	12%
China	70,000	294,000	24%
Japan	22,000	229,000	9%
Russia	43,000	203,000	21%
Canada	67,000	111,000	60%
France	21,000	110,000	19%
Germany	3,000	109,000	3%
India	25,000	108,000	23%
UK	1,000	72,000	2%
Brazil	59,000	69,000	86%
Italy	13,000	67,000	20%
Ukraine	5,000	54,000	9%
South Korea	2,000	50,000	3%
Spain	12,000	46,000	26%
Australia	6,000	43,000	14%
South Africa	1,000	40,00	—
Mexico	10,000	39,000	25%
Sweden	16,000	34,000	49%
Poland	2,000	31,000	7%
Iran	2,000	31,000	7%
Norway	27,000	27,000	99%
Turkey	11,000	26,000	40%
Taiwan	4,000	26,000	17%
Argentina	10,000	24,000	41%
Saudi Arabia	—	23,000	—
Romania	6,000	22,000	27%
Venezuela	13,000	21,000	62%

(Contd.)

TABLE 1.2 (Contd.)

(1)	(2)	(3)	(4)
Netherlands	0	21,000	0%
Indonesia	3,000	21,000	15%
Thailand	3,000	19,000	16%
Kazakhstan	2,000	17,000	13%
Pakistan	5,000	17,000	28%
Finland	3,000	16,000	18%
Switzerland	10,000	14,000	72%
Belgium	0	14,000	1%
Austria	8,000	14,000	56%
Czech Republic	1,000	14,000	7%
Egypt	3,000	13,000	21%
Colombia	9,000	13,000	65%
Malaysia	2,000	13,000	16%
Denmark	0	13,000	0%
Philippines	2,000	12,000	19%
Bulgaria	2,000	12,000	15%
Uzbekistan	2,000	12,000	15%
Hong Kong	—	11,000	—
Yugoslavia	3,000	11,000	26%
Portugal	5,000	11,000	42%
Chile	4,000	10,000	39%
Greece	2,000	10,000	23%
Others	77,000	251,000	31%
Total	694,000	3262,000	21%

Source : As on Table 1.1.

generate world electricity demand. On the basis of the study expects worldwide Solar Thermal Energy capacity of about 12,000 MW which is projected to amplify to 18,000 MW by 2020.

Geothermal Energy is commonly defined as the heat stored within the earth. The earth's temperature increases by about 3°C for every 100 meters in depth. Geothermal energy has been used for bathing and washing for thousands of years. It was only in the Twentieth century that it has been harnessed a large scale for space heating, industrial use and electricity production. Presently about 44 TWh of electricity and another 38 TWh equivalent of heat is produced commercially using this source of energy out of 12,000 TWh per year potential for electricity generation according to UNDP assessment of world geothermal potential. However, like other renewable resources like solar and wind, geothermal energy is also widely dispersed. Therefore, the technological ability, not the resource potential, will determine its future share in global energy mix.

1.2 INDIA'S HYDROELECTRIC SCENARIO

Power development in India commenced at the end of the 19th century with the commissioning of electricity supply in Darjeeling during 1897, followed by commissioning of a hydropower station at Sivasamudram in Karnataka during 1902. In the pre-Independence era, the power supply was mainly in the private sector that too restricted to the urban areas. After independence, sincere efforts have been made to develop the power sector rapidly in the country. India has the distinction of being the only country in the world to have an exclusive Ministry dealing with new and renewable energy sources[8]. With the formation of State Electricity Boards during Five-Year Plans, a significant step was taken in bringing about systematic growth of power supply industry all over the country. A number of multi-purpose projects came into being and with the setting up of thermal, hydro and nuclear power station, power generation started increasing radically.

The Ministry of Power is basically responsible for the development of electrical energy in the country. The Ministry is concerned with the perspective planning, policy formulation, processing of projects for investment decision, monitoring of the implementation of power projects training and manpower development and the administration and enactment of legislation with regard to thermal and hydropower generation, transmission and distribution. In all technical and economic matters, Ministry of Power is assisted by the *Central Electricity Authority (CEA)*.

The construction and operation of generation and transmission projects in the Central sector are entrusted to Central Sector Power Corporations viz. the *National Thermal Power Corporation (NTPC), the National Hydroelectric Power Corporation (NHPC), the North-Eastern Electric Power Corporation (NEEPCO) and the Power Grid Corporation of India Limited (PGCIL)*. The Power Grid is responsible for all the existing and future transmission projects in the Central Sector and also for the formation of the National Power Grid. Two joint venture power corporations, namely, *Satluj Jal Vidyut Nigam (SJVN) (formerly known as NJPC) and Tehri Hydro Development Corporation (THDC)* are responsible for the execution of *the Nathpa Jhakri Power Project* in Himachal Pradesh and Projects of *Tehri Hydro-Power Corporation* in Uttarakhand respectively. Three statutory bodies, i.e. the *Damodar Valley Corporation* (DVC), The Bhakra-Beas Management Board (BBMB) and Bureau of Energy Efficiency (BEE), are also under the administrative control of the Ministry of Power. Programmes of rural electrification are provided financial assistance by the Rural Electrification Corporation (REC). The Power, Finance Corporation (PFC) provides term finance to projects in the power sector. The autonomous bodies (societies), namely, Central Power Research Institute (CPRI) and the National Power Training Institute (NPTI) are also under the administrative control of the Ministry of Power. A Power Trading Corporation has also been incorporated primarily to support these Mega Power Projects in private sector by acting as a singly entity to enter into Power Purchase Agreements (PPAs).

1.2.1 Capacity Accumulation

In order to meet the projected power requirement by 2012, an additional capacity of 1,00,000 MW is required in the next two Five-Year Plans. A capacity of nearly 41,110 MW (Thermal: 25416.24 MW) have been set-up during the Tenth Plan period and the leftover would be set-up in the Eleventh Plan with a stronger focus on hydropower. The Central Sector would contribute 22,832 MW (Thermal: 12790 MW), the State Sector 11,157 MW (Thermal: 6675.64 MW) and Private Sector 7121 MW (Thermal: 5950.6 MW) in the Tenth Plan (India, 2006).

The installed power generation capacity in India has increased significantly from 1400 MW in 1947 to 1, 18,419.09 MW as on 31 March, 2005 comprising 80,902.45 MW thermal, 30,935.63 MW hydro, 38,11.01 MW wind and 2770 MW nuclear. A capacity addition programme of 6344.52 MW was fixed for the year 2005-06. Considering the fact that a large chunk of proportion of the installed capacity will come from the public sector, the outlay for the power sector was raised from Rs. 45,591 crore during the Ninth Plan to Rs. 1,43,399 crore in the Tenth Plan. This included a gross budgetary support of Rs. 25000 crore and the remaining Rs. 1,18,399 crore was from internal and extra budgetary resources. Power generation during 2004-05 was 587.366 BUs comprising 486.031 BUs thermal, 84.947 BUs hydro. The target of Power generation for 2005-06 was fixed at 621.500 BUs. The plant load factor has shown a steady improvement over the years and has improved from 52.8% in 1990-91 to 74.8% in 2004-05.

1.2.2 Hydroelectric Potential

The country has been classified into six major river systems namely Indus, Brahmputra, Ganga, Central Indian River System, East Flowing River System for the purpose of hydro electric potential survey. These river systems have been further divided into 49 basins. The first systematic and comprehensive study to assess the hydroelectric resources in the country was undertaken during the period 1953-59 by the Power Wing of the erstwhile Central Water and Power Commission. These studies placed the economical exploitable hydropower prospective of the country at 42,100 MW at 60% load factor, similar to an annual energy generation of 221 billion units.

The re-assessment studies, accomplished by Central Electricity Authority in 1987, have placed the hydro power potential at 84,044 MW at 60% load factor. A total of 845 hydroelectric schemes have been identified in the various basins, which are annually projected to yield about 442 billion units of electricity.

A vision paper highlighting comprehensive approach for development of 1,50,000 MW of Hydropower matching to the assessed potential of 84,044 MW was prepared by CEA and submitted to Ministry of Power during March 2001. This gives a road map for expediting hydro development so as to harness total potential by 2025-26. The cost of development of the remaining untapped potential in the country has been estimated to be about Rs. 5,00,000 crores at current price levels.

Further, hydroelectricity generation is not linked to issues concerning fuel supply, especially the price volatility of important fuels. It enhances our energy security and is ideal for meeting the peak demand. Its share has gradually declined. Therefore, thermal generation, which should generally be used for base load operation, is also being used to meet peaking requirements. This leads to non-optimal utilization of economic and perishable resources. Therefore, exploitation of vast hydroelectric potential has been provided extra thrust in the capacity addition plans and has been accorded high priority in our power development plans. In order to improve the thermal-hydro mix following measures have been taken by the Government of India:

- Higher budgetary allocation for the Hydel sector;
- Investment approval for new hydroelectric projects;
- Identification of new projects in the central sector for advance action;
- Promoting state sector projects which were languishing or could not progress due to inter-State disputes;
- Improving tariff dispensation for Hydel projects;
- Simplification of procedure for transfer of clearance; and
- Levy of 5% development surcharge to supplement resources for hydroelectric projects by NHPC allowed by CERC.

The hydroelectric schemes (estimated up to the year 2001) in operation are shown in Table 1.3 which account for only 17% and those under execution for around 6% of the total potential. Thus the bulk of the potential—77% remains yet to be developed. This table gives state wise assessed potential and developed and under developed. In addition to potential of medium and major hydro schemes, a sizable potential also exists for development of micro, mini and small hydro schemes

TABLE 1.3
State-wise Status of Hydro Potential Development

State	*Potential assessed at 60% LF (MW)*	*Potential Developed (MW)*	*% Developed*	*Potential Under Development*	*% Under Development*
(1)	(2)	(3)	(4)	(5)	(6)
Jammu & Kashmir	7487	508	6.8	379	5.1
Himachal Pradesh	11647	2037	17.5	608	5.2
Punjab	922	656	71.2	173	18.8
Haryana	64	52	80.7	12	18.2
Rajasthan	291	193	66.2	8	2.7
Uttarakhand	7453	832	11.2	1326	17.8
Madhya Pradesh	2774	600	21.6	1203	43.4
Gujarat	409	139	33.9	111	27.1
Maharashtra	2460	1119	45.5	187	7.6
Andhra Pradesh	2909	1402	48.2	34	1.2
Karnataka	4347	2323	53.4	310	7.1
Kerala	2301	1126	48.9	219	9.5
Tamil Nadu	1206	947	78.5	68	5.6
Jharkhand	478	75	15.7	211	44.1
Bihar	60	45	74.6	0	0.0
Orissa	1983	1101	55.5	9	0.5

West Bengal	1786	91	5.1	10	0.6
Sikkim	1283	58	4.5	109	8.5
Meghalaya	1070	122	11.4	0	0.0
Tripura	9	9	94.4	0	0.0
Manipur	1176	73	6.2	48	4.1
Assam	351	112	31.8	91	25.9
Nagaland	1040	56	5.4	26	2.5
Arunachal Pradesh	26756	17	0.1	108	0.4
Mizoram	1455	1	0.1	37	2.5
All India	84,044	14,003	16.7	5294	6.35%

Source : As on Table 1.1, p. 301.

Fig. 1.2
State-wise Hydro Potential Development

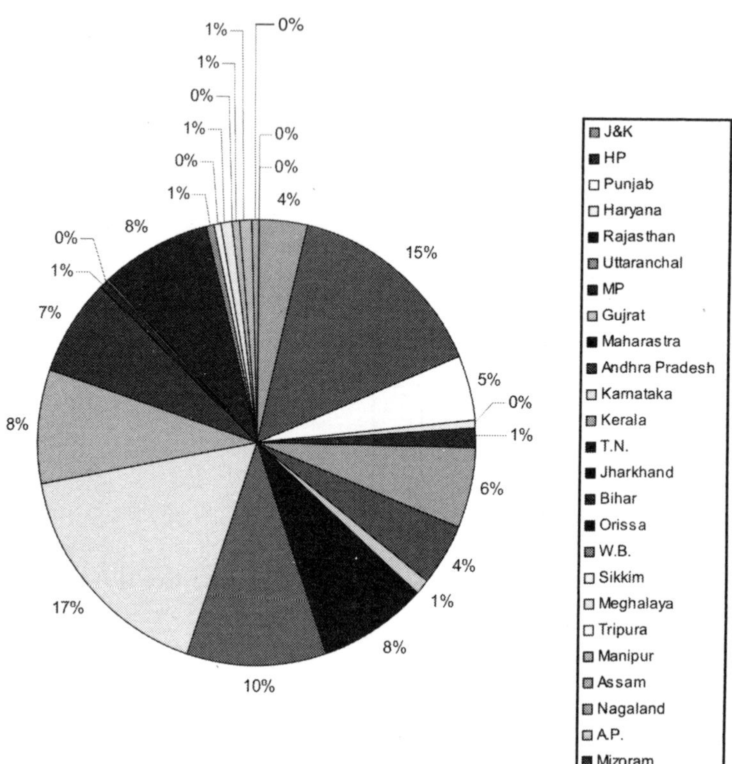

Based on Table 1.3.

on rivulets and canal drops. However, Government of India is eagerly trying to tap available hydroelectric potential of the country. Consequently, from the Eleventh plan, the budgetary allocation for hydroelectric development has given special emphasis in view of rapid industrialization and modern economic growth.

1.2.3 Major Hydropower Projects

The importance of physical infrastructure, particularly power for sustained economic development is well recognized. The physical infrastructure through its backward and forward linkages facilitates growth. In order to accelerate to pace of development, sincere effort for hydroelectric development (being most economical) is required. Initially the major hydroelectric power projects were commissioned under public sector. The descriptions of some of the major hydropower projects are given below:

1.2.4 National Hydroelectric Power Corporation Limited

The National Hydroelectric Power Corporation Limited,[9] is a Schedule 'A' Enterprise of the Government of India with an authorized share capital of Rs. 1,50,000 million and an investment base of over Rs. 2,22,000 million. Established in 1975, in its existence of about 31 years, NHPC has become the major organization for hydropower development in India, with capabilities to undertake all the activities from "concept to commissioning" in relation to setting up of hydro projects. The NHPC has ISO-9001 certification for its quality management system and ISO-14001 for environment standard in Corporate Office. The NHPC has completed 9 hydro electric projects with installed capacity of 2,755 MW on its own and one project, i.e., Indira Sagar (1,000 MW) in Madhya Pradesh which is in Joint Venture with Government of M.P. 10 projects of NHPC with installed capacity of 5,103 MW and one project in JV with Government of MP, i.e., Omkareshwar (520 MW) is under active construction stage. In addition NHPC has also completed 14.1 MW Devighat Hydropower Project in Nepal, 60 MW Kurichu Hydro Power Project in Bhutan and 5.25 MW Kalpong Hydro Power Project.

The NHPC is a profit-earning organization since becoming operational with profit showing an upward trend. The NHPC power stations generated highest ever generation of 12,567 million units and achieved Capacity Index of 98.16 per cent during 2005-06. NHPC has been assigned the highest rating 'AAA' (ind) by Fitch ratings for domestic borrowings and Sovereign rating for external borrowings by Fitch ratings and S&P. The NHPC has been providing consultancy services in the area of hydro power development to the clients in India and abroad. NHPC is registered as Consultant in the area of Hydro Power with International Funding Agencies like World Bank, Asian Development Bank, African Development Bank, etc. New

Consultancy assignments amounting to Rs. 65 crore were received during the financial year 2004-05. The NHPC has signed a Memorandum of Understanding with Uttarakhand Government for implementation of 240 MW Chungar Chal, 630 MW Garba Tawaghat and 55 MW Karmoli Lumti Tulli Projects in Uttarakhand. Power Purchase agreements have been signed for Kishanganga, Nimmo Bazgo, Chutak, Uri-II, Dul Hasti, Chamera-III and Teesta Low Dam Project Stage-IV with the concerned beneficiaries. The NHPC has signed agreements with Government of Sikkim for execution of the 495 MW Teesta Stage-IV and 210 MW Lachen Hydroelectric Projects in Sikkim on BOOM basis. The Organization has launched a major initiative to implement Enterprise Resource Planning (ERP) software solution framework for integrating major business processes to achieve quality of work, improved profitability and performance.

1.2.5 Ratnagiri Gas and Power Private Limited

Ratnagiri Gas and Power Private Limited was formed on 08 July 2005 as a joint venture between NTPC, GAIL, MSEB holding company and Indian Financial Institutions to takeover the assets of the erstwhile Dabhol Power Company Limited, complete the project and operationalise it. NTPC has invested Rs. 500 crore as 28.33 per cent equity in this project. The Block-II (740 MW) of the 2150 MW Dabhol project was taken up for revival on fast track basis after transfer of its assets in October 2005 to RGPPL and has already been synchronised in April-May 2006.

The NTPC is implementing its first hydro project, 800 MW Koldam Hydroelectric Power Project (HEPP), Himachal Pradesh. Further, NTPC has signed the Implementation Agreements for execution of Loharinag Pala HEPP (600 MW), Tapovan-Vishnugad HEPP (520 MW) and Rupsiabagar Khasiyabara HEPP (260 MW. Moreover, NTPCs subsidiary NTPC Hydro Ltd. (NHL) has signed the Implementation Agreements for execution of Lata-Tapovan (171 MW) HEPP and Rammam-III (120 MW) HEPP. During the year 2005-06, a consortium comprising NTPC Ltd., Canoro Resources Ltd. and Geo Petrol International has been allotted an oil exploration block in Energy 267 Arunachal Pradesh. A Production Sharing Contract (PSC) for the block has been signed between the Government of India and the Consortium. The NTPC was allotted Pakri Barwadih coal mining blocks by Government of India and action have been initiated by NTPC for its development. Further, NTPC has been allotted 7 more captive coal mining blocks by the Government. These blocks are expected to produce 50 million tonnes of coal per annum. Of these 2 mine blocks at Brahmini and Chichro Patsimal in Orissa are to be developed through a 50:50 Joint Venture between NTPC and Coal India Limited. As part of its diversification drives, NTPC has formed several Joint Venture Companies such as : NTPC–SAIL Power Supply Company (P) Ltd. (NSPSCL) for operating the Captive Power Plants of Durgapur and Rourkela Steel Plants having total capacity of 240 MW Bhilai Electric Supply Company Ltd. (BESCL),for operating Captive Power plant (74 MW) at Bhilai Steel Plant of SAIL. The Company is also implementing 500 MW (2x250 MW) expansion of Bhilai Captive Power Project. NTPC Alstom Power Services Limited (NASL) for taking up Renovation and Modernisation (R&M) assignments of power plants in India and abroad; Utility Powertech Limited (UPL) for taking up assignments of construction, erection and project management in power and other sectors; NTPC Tamil Nadu Energy Company Limited (NTECL) to set-up a coal-based power station of 1000 MW capacity, at Ennore in Tamil Nadu, using Ennore Port infrastructure facilities, Ratnagiri Gas and Power Private Ltd. (RGPPL) for revival/restart of Ratnagiri Project (erstwhile Dabhol Project). NTPC has also formed several wholly owned subsidiaries such as : NTPC Vidyut Vyapar Nigam Limited (NVVN) for trading in Power. NVVN is also engaged in facilitating development of power exchange in India; NTPC Electric Supply Company Ltd. (NESCL) to take up power distribution business; NTPC Hydro Limited (NHL) for development of small and medium hydropower projects of capacity less than 250 MW; Pipavav Power Development Company Ltd. (PPDCL) for processing of setting up a 1000 MW thermal power project at Pipavav. NTPC has extended benefits of its success and experience by providing services for the development of Indian Power Sector. NTPC has participated in several programmes of Subsidiaries, such as: (i) "Accelerated Rural Electrification Programme" (AREP AREP), later merged with the programme of Electrification of Rural Villages & Households under the name of "Rajiv Gandhi Grameen Vidyutikaran Yojana (RGGVY)", under which NTPC is involved in turnkey execution of rural electrification work in 40,000 villages in 6 states. (ii) Distribution Reforms and Upgradation Management (DRUM)—a collaborative effort of Ministry of Power and USAID, with NTPC Power Management Institute (PMI) as a leading partner in delivering training programmes on distribution business. (iii) APDRP (Distribution)—NTPC has been identified as the Lead Advisor-*cum*-Consultant (AcC) for implementation of APDRP programme targeted at improvement of distribution sector in 12 states, out of which 6 states are directly assigned to NTPC and balance 6 are through field AcC-CPRI and MECON. (iv) Partnership in Excellence—CEA has identified 26 stations which are operating at a PLF of less than 40 per cent. These stations would have a "Partnership in Excellence" with better performing utilities, so as to achieve an improvement in performance in the shortest possible time. NTPC has been entrusted the responsibility of 14 stations out of the identified 26. (v) R&M of SEB stations—The 10th Plan envisages R&M of an installed capacity of 11.055 MW with a target of achieving 75-80 per cent PLF and 20 years of life extension. NTPC has taken up consultancy services for a few of these stations.

1.2.6 Power Grid Corporation of India Limited

The Power Grid Corporation of India Limited (PGCIL)[10] was included as a Government enterprise on 23 October 1989 for organization of regional and national power grids to facilitate transfer of power within and across the regions with reliability, security and economy and on sound commercial principles. It has been known as a mini-ratna category-I PSU. Presently, about 45 per cent of total power generated in the country is being transferred over PGCIL's transmission network. During the year 2005-06, the Organisation has earned a net profit of about Rs. 915 crore (Provisional) on a turnover of about Rs. 3,245 crore (Provisional) and commissioned about 4,400 ckt. kms. of transmission lines and 8 new sub-stations and added transformation capacity of about 5,000 MVA. With a view to operate, monitor and control the Regional Power Grids in a unified, well coordinated and integrated manner, PGCIL has established Unified Load Dispatch and Communication (ULDC) schemes in all the regional power grid of India, namely, Northern, Southern, North-Eastern, Eastern and Western regions. Further, to facilitate smooth operation of National Grid on real time basis and free uninhibited bulk exchange of power among the regions, National Load Dispatch Centre at Delhi is being set-up. PGCIL has diversified into Telecom business to utilize spare telecommunication capacity available with its Unified Load Dispatch Centre (ULDC) schemes and leveraging its country wide transmission infrastructure. It envisages establishing telecom network of about 20,000 kms interconnecting about 60 major cities including Metros and all State capitals at a cost of about Rs. 1,000 crore. PGCIL has been able to commission about 18,700 kms of network since approval was accorded by Government in March 2003. The balance network is in an advanced stage of completion.

1.2.7 Satluj Jal Vidyut Nigam Limited

The successful completion and commissioning of the prestigious 1500 MW Nathpa Jhakri Hydro Power Project in Himachal Pradesh in September 2003 was an important landmark for the Corporation. The Project has already generated 12,386 million units (MU) of electrical energy up to 20 June 2006.

Nathpa Jhakri Hydro Electric Project was dedicated to the nation by the Hon'ble Prime Minister of India Dr. Manmohan Singh on 28 May, 2005 in the august presence of Hon'ble Union Minister of Power at a function presided over by Hon'ble Chief Minister of Himachal Pradesh. After successful implementation of this Project, SJVN has embarked on a target of becoming 5000 MW multi-project company. It has already taken up six more projects in Himachal Pradesh and Uttarakhand

The Satluj Jal Vidyut Nigam Limited, SJVN (formerly Nathpa Jhakri Power Corporation Limited (NJPC) was incorporated on 24 May 1988 as a joint venture of the Government of India and the Government of Himachal Pradesh (GOHP) with an authorized share capital of Rs. 4500 crore. The debt equity ratio for the Nathpa Jhakri Hydro Electric Project

(NJHEP) is 1:1 and the equity—sharing ratio of GOI and GOHP is 3:1 respectively.

Besides the social and economic development of the people in its vicinity, the 1500 MW NJHEP is designed to generate 6950 MW of electrical energy, in a 90 per cent dependable year. It also provides 1500 MW of valuable peaking power to the Northern Grid. Out of the total energy generated at the Bus Bar, 12 per cent is supplied free of cost to the home state, Himachal Pradesh From the remaining 88 percent energy generation 25 per cent is supplied to HP at Bus bar rates. The Corporation has bagged a number of awards for environment protection safety and eco-systems. It also has the ISO 9001 certification.

SJVN has the experience of corporate and project planning, design, engineering, construction management, erection and commissioning, contracts management, project management, human resource management, financial management and commercial management of India's largest hydroelectric project. To effectively utilize the in-house expertise and the experience gained, a dedicated consultancy division has been established for providing consultancy services to national and international organizations.

1.2.8 Tehri Hydro Development Corporation Limited

THDC, a Joint Venture Corporation of the Government of India and Government of U.P., was incorporated in July 1988 as a Limited Company under the Companies Act, 1956, to develop, operate and maintain the Tehri Hydro Power Complex and other Hydro Projects. The Corporation has an authorized share capital of Rs. 4000 crore. THDC is presently responsible for the implementation of the Tehri Hydro Power Complex (2400 MW), on the river Bhagirathi, comprising of Tehri Dam & Hydro Power Plant (HPP) (1000 MW), Koteshwar Hydro Electric Project (HEP) (400 MW) and Tehri Pumped Storage Plant (PSP) (1000 MW). The Tehri Stage-I is at an advance stage of completion. Two units of 250 MW each were successfully rolled on 31 March 2006. With this, commissioning process of Tehri Stage-I Project has started. Generation from the Project was planned to commence from June 2006 onwards. The works of Koteshwar HEP are in progress and the project is scheduled to be commissioned by March 2008. Essential works of Tehri PSP has been taken up along with the execution of Tehri Stage-I. Investment approval of Tehri PSP is under process. The Project is envisaged to be completed in 4 years after investment approval.

The Tehri Hydro Power complex (2400 MW) will generate 6200 Million Units of energy annually on its completion (3568 Million Units on completion of Tehri Stage-I) and will provide additional irrigation facility in 2.70 lakh hectares of land besides stabilising existing irrigation facility in 6.04 lakh hectares of land. The project will provide drinking water for 40 lakh people in Delhi and for 30 lakh people in the town and villages of Uttar Pradesh.

The Corporation is also entrusted with the Vishnugad Pipalkoti Project (444 MW) on river Alaknanda, Kishau Dam and HPP (600 MW) on river Tons (tributary of Yamuna), and 6 other new Hydro Projects in Uttaranchal, totaling to 695 MW. The Government of Uttaranchal (GOUA) has signed an MoU for preparation of Detailed Project Report (DPR) of Vishnugad Pipalkoti Project in April 2003. The Stage-I activities of the project are already complete. DPR envisaging an installed capacity of 444 MW (4×111) has been submitted to Central Electricity Authority in March 2006 for Techno-Economic Clearance. The Project is scheduled for commissioning by Year 2011, i.e., during Eleventh Plan. The Government of Uttaranchal (GOUA) also allocated following Six Projects to THDC for development:

Name of Projects	*Capacity*	*River*	*District*
Karmoli	140 MW	Jadhganga	Uttarkashi
Gohana Tal	60 MW	Birahiganga	Chamoli
Jadhganga	50 MW	Jadhganga	Uttarkashi
Maleri Jhelam	55 MW	Dhauliganga	Chamoli
Jhelam Tamar	60 MW	Dhauliganga	Chamoli
Bokang Bailing	330 MW	Dhauliganga	Phithoragarh

Source : India 2006, Government of India, p. 270.

Implementation Agreement between THDC and GOUA for above new Projects were signed in November 2005. Investment approval for carrying out Stage-I activities for five Projects excluding Bokang Bailing has been accorded by Government of India in March 2006 and works for Stage-I activities have been taken up. The Government of Uttaranchal (GOUA) has conveyed "in principle" approval for allocation of Kishau Dam (600 MW) to THDC. The project will generate 1216 MU of power annually, besides providing irrigation (97,076 ha.) and drinking water to Delhi (616 MCM). THDC would be taking up the project once issue of appointment of cost among Power, Drinking Water Supply and Irrigation Components are firmed up and Implementation Agreement/MOU is signed.

1.2.9 Bhakra Beas Management Board[11]

The Bhakra Beas Management Board (BBMB) manages the facilities created for harnessing the waters impounded at Bhakra and Pong in addition to those diverted at Pandoh through the BSL Water Conductor System. It was also assigned the responsibility of delivering water and power to the beneficiary states in accordance with their entitled shares. The Board is responsible for the administration, maintenance and operation at Bhakra Nangal Projects, Beas Project Unit I and Unit II including Power Houses and a network of transmission lines and grid sub-stations. The Power generation of BBMB power evacuation system running into 3735 circuit km length of 400 KV, 220 KV, 132 KV and 66 KV transmission lines and 24 EHV sub-stations. The installed capacity of BBMB Power plants is 2,866.30 MW. The generation during 2005-06 was 11692 MU.

1.2.10 Damodar Valley Corporation

The Damodar Valley Corporation (DVC), the first multipurpose river valley project of the Government was set up in July 1948 for the unified development of Damodar Valley region spread over the States of Jharkhand and West Bengal. It's objectives include flood control and irrigation, water supply and drainage, generation, transmission and distribution of electrical energy, both hydroelectric and thermal, afforestation and control of soil erosion, public health and agricultural, industrial, economic and general well-being in the Damodar Valley. The DVC's main projects include four dams at Maithon, Panchet, Tilaiya and Konar with connected hydroelectric power stations (except at Konar), thermal power stations at Bokaro 'A', Bokaro 'B', Chandrapura, Durgapur, Mejia and also one gas turbine station at Maithon. The total existing capacity of DVC power plants, as on April 2006, is 2931.5 MW comprising of 2705 MW thermal, 144 MW hydel and 82.5 MW from gas turbine station. DVC's transmission network (220 KV and 132 KV) runs to a total length of 4734 circuit kms. The T&D system of DVC is supported by 42 sub-stations, 15 receiving stations and 964 circuit kms. of distribution lines. DVC supplies power at voltage levels of 33 KV, 132 KV and 220 KV to the core sector industries in the region viz. coal mines, steel plants, railways and other big/medium industries and also to its licensees including States Electricity Boards of Jharkhand and West Bengal. As a constituent of Eastern Regional Grid, DVC is also exporting its surplus power to the deficit regions of the country through the Central Transmission Utility network.

As per the programme of 1210 MW capacity addition by DVC during the Tenth Five-Year Plan, addition of 210 MW has already been achieved in 2004-05 and construction works for another 1000 MW are in progress for commissioning during 2006-07. In addition, DVC has also taken up action for refurbishment of its old thermal and Hydel units through RLA-based R&M/LE. Matching extension, augmentation and strengthening of its transmission and distribution network has also been undertaken by the Corporation.

1.3 GOVERNMENTAL INITIATIVES TOWARDS ACCELERATED HYDROPOWER DEVELOPMENT

1.3.1 Development of Private Small Hydropower Plants

In India, power generation though a Hydel resource is presently categorized as micro Hydel up to 100 kW capacity, mini Hydel up to 2000 kW and small Hydel upto 25000 kW capacity. The water resources capable of power generation in these ranges have largely remained untapped in India, since the focus has been primarily on large scale power generation.

Considering the benefits of Small Hydropower (SHP) projects, a movement has started to speedily develop such projects. As a result of this small hydro projects in Asia Pacific region have grown to a significant number. China in particular dominates in scene where more than 41,000 small hydropower stations (up to 25 MW) with an installed capacity of 25,600 MW have been built by 2004 having annual energy output of 87 billion kWh covering almost 330 million populations.

1.3.2 Potential and Policies

India has an estimated potential of more than 15,000 MW of small hydropower in natural streams, canal falls and dam toes. Most of the opportunities are available in Himalayan and other hilly regions, where fast flowing and perennial streams can easily be exploited to tape this renewable source of energy. In the past our country laid emphasis on large coal, gas and nuclear power stations. However, these large power stations posed many environmental and ecological problems.

In view of the above, Government of India started giving due attention towards the small hydropower projects. These small hydropower stations are technically and economically comparable with thermal, diesel or large hydropower stations, besides being environment-friendly with low gestation period and require little input for operation and maintenance. SHP projects up to Rs. 100 crores are exempted from environmental clearance. To bridge the gap between the demand and supply of Electricity, Government of India and many state governments has thrown open the Small Hydro Sector to private enterprises.

With a view to promote and harness the exploitable small hydro power potential of around 15,000 MW capacity, the responsibility of SHP development (up to 25 MW) lies with Ministry of Non-Conventional Energy Sources. After consultation with the concerned departments in the Central Government and the State Government, has finalized the strategy and the incentives required for promoting SHP projects of capacity up to 25 MW each. The new strategy envisages the development of SHP mostly through commercial projects with private sector participation. To encourage the private sector, Ministry of Non-Conventional Energy Sources (MNES) announced incentives for projects up to 25 MW capacities. The incentive schemes, provides financial support for conducting feasibility studies, capital subsidy for commercial projects and capital grant to the state sector projects. A higher level of incentives has been provided for the north-east states and the hilly regions to exploit their available potential.

These incentives for small hydropower projects by MNES and respective State Governments have generated enthusiasm among the private promoters. Small hydropower development has provided unique opportunity for profitable investment. In fact, SHP developers in Himachal Pradesh, Uttarakhand, Kerala, Karnataka, Madhya Pradesh and Andhra Pradesh, etc. have gone ahead and signed agreements with respective State Governments. Many private SHP developers in Karnataka, Andhra Pradesh, Uttarakhand, Himachal Pradesh and other states are generating and supplying power to grid through SHP projects.

Small hydro technology is mature and proven. Construction is straight forward and involves simple processes, which offer opportunities for a large degree of local participation, both in terms of labour and materials. Construction lead times are short. Established and proven design concepts offer considerable scope for adaptation to local circumstances, both in construction and operation and may range from simple manual attention to fully automatic and computerized system. Small hydro developments are usually run-of-river development where water is used only as it is available, and with no water storage reservoir. The cost of large dams can rarely be justified for small projects. Therefore, a low dam or diversion weir of simplest construction is usually built.

Small projects[12] lack the advantage of scale and their cost per KW installed can therefore be quite high. Present investment costs for projects in the range of 2000 kW to 25,000 kW can be expected to lie in a range of about Rs. 40,000-60,000 per kW installed. Higher costs are likely to be justifiable only in special case. Lower costs might be experienced at particularly favorable sites or where there is an appreciable amount of lower cost local input. Project costs per kW usually decrease with increasing capacity and increasing head but design parameters are normally defined by local conditions and allow little latitude for change. As it is often quite inexpensive to add a power component to an existing water supply or irrigation scheme, multi-purpose projects could well provide one of the platforms for the expansion of small hydro in the future.

The smaller the scheme the greater the need to concentrate on capital cost reduction. Non-essential features are eliminated and local materials are used whenever possible. Selections of the appropriated control equipment require a compromise between complexity and cost and the choice may be constrained by available skills. Achieving satisfactory performance with simple manual control requires relatively high skill by the operator. A high degree of automation, on the other hand, puts the emphasis on adequate maintenance and the procurement of spares, which in turn could entail greater reliance.

Big and medium size hydro schemes may create environmental impacts of various kinds. Such impacts tend to be highly site-specific and to be influenced by the civil design adopted. Some loss of habitat may occur due to the need for access roads, transmission lines, and the loss of habitat may occur due to the need for access roads, transmission lines, and the construction of the structure. Impacts of this type may be counterbalanced by improvements in resource utilization, recreational or other social benefits. Where construction of reservoirs or enhancement of natural storage is involved, water quality may be impacted, salutation may occur and the character or river flow may be altered. Water quality may be affected by leaching of pollutants from newly inundated land, increased bacterial oxygen demand and stratification.

Small hydro does not usually involve the construction of significant storage and in general avoids significant impacts of the types noted above. However, the diversion of water from natural channels and its passage through a turbine in inevitable and may create due to fish mortality of impediments to the movement of migratory fish species. In general, the environmental impacts noted above are not significant with small hydro projects, certainly relative to the benefits of the projects or to alternative means of supply electricity.

The present scenario of private small hydropower development is encouraging in southern states where many private small hydropower projects are supplying power to regional grid and big consumers. In northern states like Uttranchal and Himachal Pradesh, the developers are finding many roadblocks and difficulties in project approvals, power purchase agreement, land acquisition and statutory clearance.

Only few SHP projects allotted to private sector in U.P., Uttranchal and Himachal Pradesh are implemented. Various issues related to PPAs, such as tariff, deemed generation, letter of credit, escrow account; state government guarantee power evacuation arrangement, etc. on which agreement was signed had been renegotiated by the state government making these project developers difficult to tie up the financing. With little success story of private SHP the project developers and the state governments are moving slowly looking helplessly the enormous energy going waste.

Under Accelerated Power Development and Reforms Programme (APDRP), PGCIL has been assigned the role of Advisor-*cum*-consultant to lend its managerial and technical expertise for developing 182 distribution circles/town/schemes spread over 18 States costing about Rs. 7,820 crore. The Company is also executing APDRP schemes of about Rs. 1,100 crore on deposit work basis under bilateral arrangement on behalf of States like Goa, Bihar, Meghalaya, Uttar Pradesh, Tripura and Gujarat. During the year 2005-06, PGCIL electrified 2,682 villages in Bihar, Uttar Pradesh and West Bengal against a target of 2,625 villages. Quadripartite Agreements were signed amongst State governments, State Utilities, PGCIL and Rural Electrification Corporation (REC) for implementation of Rajiv Gandhi Grameen Vidhyutikaran Yojana (RGGVY) Schemes in 9 (Nine) States involving approx., 87,300 Villages of 68 Districts at an estimated cost of Rs. 9,400 crore.

The Power Finance Corporation Limited was incorporated on 16 July, 1986 and registered as a Non-Banking Financial Institution by Reserve Bank of India in February 1997. Set-up as a Power Sector Financial Institution, the Company has been a dominant player in funding the projects of the State Power Utilities. Over the last 4 to 5 years, the Company has also started playing a major role in funding of Central Sector and Private Power Projects. The Company's loan portfolio of Rs. 35,603 crore as on 31 March 2006 includes various types of Power Projects viz. Generation, Transmission, Renovation and Modernisation, System Improvement, etc. In Generation segment, the projects funded by the Company include coal-based thermal, gas-based thermal, hydro, bio-mass and wind power projects. The company has recently been identified as the nodal agency for development of presently seven number of Ultra Mega Power Projects of 4000 MW each in various parts of the country through tariff-based competitive bidding. These include three pit-head-based projects which would include development of captive coal mines and four coastal projects based on imported coal. Each of these projects will be progressively commissioned over the next 5 to 7 years. The authorized share of the Company is Rs. 2000 crore and paid up capital is Rs. 1030.45 crore. PFC has cumulatively sanctioned Rs. 94,052 crore and disbursed Rs. 61,799 crore up to 31 March 2006. During the financial year 2005-06, the Company has sanctioned Rs. 22,502 crore and disbursed Rs. 11,681 crore. The post-tax profit is Rs. 970.95 crore for 2005-06. The Company is continuously rated "Excellent" since its signing MoU with Government of India and received 'MOU Award of Excellence' for FY 2003-04 for being among the 'Top 10 Public Sector Undertakings of the country' from the Hon'ble Vice President of India.

The Rural Electrification Corporation Limited (REC)[13] was incorporated as a Company under Companies Act, 1956 in 1969 with the main objective of financing rural electrification schemes in the country. The current mission of REC is to facilitate availability of electricity for accelerated growth and for enrichment of quality of life of rural and semi-urban population and to act as a competitive, client-friendly and development-oriented organization for financing and promoting projects covering power generation, power conservation, power transmission and power distribution network in the country.

With a need to develop the huge power potential North Eastern Electric Power Corporation (NEEPCO) was incorporated on 2 April 1976 as a wholly owned Government Enterprise under the Ministry of Power to Plan, Promote, Investigate, Survey, Design, Construct, Generate, Operate and Maintain power stations in the N.E. Region. The authorized share capital of the Corporation presently stands for Rs. 3,500 crore. The installed capacity of NEEPCO is 1,130 MW comprising of 755 MW of Hydropower and 375 MW of gas based power. The Corporation currently meets more than 68 per cent of the energy requirement of the North-Eastern Region. It is an ISO 9001:2000 (Quality) ISO 14001:1996 (environment) and OHSAS 18001:1999 (Safety) Company with its Corporate Office at Shillong. The Corporation plans to add 1380 MW during eleventh five-year plan to its present installed capacity of 1130 MW. Presently two projects of total capacity of 660 MW are under execution by NEEPCO. During the year 2005-06 NEEPCO generated 5,260 MU of energy which is 7.26 per cent more than targeted with a break up of 2758 MU from Hydro sector and 2,146 MU in Thermal. The corporation's Sales and receipt were Rs. 838.00 crore and Rs.934.00 crore respectively for the concluding year 2005-06 which is 5 per cent and 21 per cent more than the preceding year. The corporation's Net Profit during 2005-06 is Rs. 204 crore. The corporation already prepared and submitted four DPRs to CEA during 2005-06 and three Projects were taken up for preparation of DPRs during 2006-07 for execution in phases keeping in view the 50,000 MW Hydroelectric initiative launched by the Government of India.

The National Power Training Institute,[14] a registered society, an ISO 9001:2000 ISO 14001 Organization under the Ministry of Power, is committed to the development of Human Resources in Indian Power and Energy Sectors. NPTI operates on all India basis with its Corporate Office at Faridabad and the Regional Institutes located at New Delhi, Nagpur, Durgapur, Neyveli, Bangalore and Guwahati. NPTI's Corporate Centre and its Institutes are well equipped with world-class hi-tech infrastructural facilities in conducting different courses on technical as well as management subjects. Since its inception, NPTI has shared its engineering and technological expertise through its long-term and short-term training programme imparting training to more than 1,11,000 power professionals. Also through its mass awareness programmes in Energy Conservation, Power Reforms, Electrical Safety, Energy Environmental Linkage, Water for Sustainable Development of Power, etc. over 1,47,000 persons were sensitized across the country.

The Central Power Research Institute (CPRI)[15] a society registered under the societies Registration Act under the Ministry of Power serves as a national laboratory to carry out applied research in Electrical Power Engineering. It also functions as an independent National Testing and Certification Authority for Electrical Equipment for ensuring their reliability. Over the years, CPRI has built up expertise in the areas of transmission and distribution systems, power quality, energy metering, energy auditing, transmission line, tower design, conductor vibration studies, power systems instrumentation, transformer oil reclamation and testing, diagnostic studies, condition monitoring and estimation of remaining life of equipment, new material for power system application, UHV testing-short circuit testing, HV testing, seismic qualification of equipment and other related fields. CPRI offers consultancy services in these areas. The Institute works as a nodal agency for national level power system research. Among the new ventures in CPRI, Centre for Collaborative and Advanced Research (CCAR) has been established for creating infrastructure for the visiting Scientists/Technologists to carry out research in the areas of power sector. The other important facility currently added is for showcasing of all available technologies for Industrial Solid Waste Utilisation. The Institute is creating testing facilities at Kolkata and Guwahati to cater to the testing requirements in the Eastern and North-Eastern States of the country, with the infrastructure assistance from WBSEB and ASEB. The other new facility being established includes a labeling laboratory for air conditioners and refrigerators. CPRI is expanding its services to other Asian and African countries and is executing many consultancies and testing jobs.

1.3.3 State Electricity Regulatory Commissions

Many State Electricity Regulatory Commissions (SERCs) have been established under the provisions of the ERC Act, 1998 or under respective State Reforms Acts. These SERCs have been continued under the provisions of Electricity Act, 2003. Orissa, Haryana, Andhra Pradesh, Uttar Pradesh,Uttaranchal, Karnataka, Rajasthan, Madhya Pradesh, Delhi and Gujarat have enacted their State Electricity Reforms Acts, which provide, *inter-alia*, for unbundling/corporatisation of SEBs, setting up of SERCs, etc. The SEBs of Orissa, Haryana, Andhra Pradesh, Karnataka, Uttar Pradesh, Uttarakhand, Rajasthan, Delhi, Assam, Tripura, Gujarat, Madhya Pradesh and Maharashtra have been unbundled. Distribution was privatised in Orissa and Delhi. So far twenty-five states viz., Orissa, Haryana, Andhra Pradesh, Uttar Pradesh, Karnataka, West Bengal, Tamil Nadu, Punjab, Delhi, Gujarat, Madhya Pradesh, Maharashtra, Rajasthan, Himachal Pradesh, Assam, Chhattisgarh,

Uttarakhand, Goa, Bihar, Jharkhand, Kerala, Tripura, Sikkim, Jammu & Kashmir and Meghalaya have either constituted or notified the constitution of SERCs. Joint Electricity Regulatory Commission (JERC) has been notified for Mizoram and Manipur. JERC has also been notified for Union Territories (except Delhi). Twenty SERCs viz. Orissa, Andhra Pradesh, Uttar Pradesh, Maharashtra, Gujarat, Haryana, Karnataka, Rajasthan, Delhi, Madhya Pradesh, Himachal Pradesh, West Bengal, Punjab, Tamil Nadu, Assam, Uttarakhand, Jharkhand, Kerala, Chhattisgarh and Tripura have issued tariff orders.

1.3.4 Reorganization of State Electricity Boards (SEBs)

The Central Government has agreed for continuation of the SEBs of 11 States after considering their respective requests in this regard up to 9 June 2006 except for Bihar for which extension has been given up to 9 September 2006. Out of the eleven states, Madhya Pradesh and Assam have carried out the re-organization except for the area of trading of electricity. The Appellate Tribunal for Electricity established by the Central Government under Section 110 of the Electricity Act, 2003 has been operationalised. The headquarters of the Appellate Tribunal is at Delhi. The Tribunal has started hearing appeals against orders of the Regulatory Commissions/Adjudicating Officers.

The Central Government has notified Electricity Rules, 2005 on 8 June 2005 which carry provisions related to Captive Generating Plants, Consumer Redressal Forum; and Tariff of Generating Companies, etc. The Tariff Policy has been notified by Government of India under the provisions of section 3 of the Electricity Act, 2003. The objectives of the tariff policy are to : (a) Ensure availability of electricity to consumers at reasonable and competitive rates; (b) Ensure financial viability of the sector and attract investments; (c) Promote transparency, consistency and predictability in regulatory approaches across jurisdictions and minimize Perceptions of regulatory risks; and (d) Promote competition, efficiency in operations and improvement in quality of supply.

In compliance with section 63 of the Electricity Act, 2003, the Central Government has notified guidelines for procurement of power by Distribution Licensees through competitive bidding. The Central Government has also issued the standard bid document containing RFQ, RFP and model PPA for long-term procurement of power from projects having specified site and location.

The Forum of Regulators has been constituted under sections 166(2) and (3) of the Electricity Act, 2003. The Forum shall discharge, *inter-alia,* functions viz. analysis of tariff orders and other orders of CERC and SERCs and compilation of data arising out of the said orders, highlighting, especially the efficiency improvements of the utilities, laying of standards of performance of licensees, evolving measures for protection of interest of consumers and promotion of efficiency, economy and competition in power sector. Regional Power Committees have been constituted under the provision of section 2(55) of the Act for facilitating integrated operation of the power system in respective regions on 29 November 2005.

1.4 HYDROPOWER SCENARIO IN THE NORTH-WESTERN HIMALAYAN REGION (HIMACHAL PRADESH)

Himachal Pradesh has an enormous hydel potential and in the course of preliminary hydrological, topographical and geological investigations, it has been projected that about 21,244 MW of hydelpower can be generated in the State, by constructing various major, medium, small and micro-hydel projects on five perennial river basins. In addition, a large number of unidentified areas have still been left in the river basins which can contribute substantially to the power potential of Himachal Pradesh. Also in view of the rising cost of thermal and nuclear power generation, many identified projects which have been excluded due to high cost of generation, will also become viable in future. Out of the total hydel potential only 6060.00 MW has been harnessed so far, out of which only 329.50 MW is under the control of Himachal Pradesh as bulk of the potential has been exploited by the Central Government and other agencies. The huge hydel potential of the State can play a major role in power development programmes in the northern region which will provide an economic base for the overall development of Himachal Pradesh.

When Himachal Pradesh came into being, it had only one power house at Jogindernagar, which was at that time generating about 10,000 KW of power. Till the year 1964, the electricity wing was part of the Public Works Department. It was only after the visit of Dr. K.L. Rao, the then Union minister for Irrigation and Power, to Himachal Pradesh in 1964, that the idea of exploring vast hydel power potential got fructified and consequently with the active interest of Himachal Pradesh Government, the department of multipurpose projects and power was created for survey of power potential and to carry out other allied functions for the final execution of such projects. In 1966, at the time of merging the Punjab Hill areas with Himachal Pradesh, the Transmission and Distribution system located in the merged area was also transferred along with Bassi Project to Himachal Pradesh. In April 1971, Himachal Pradesh State Electricity Board was created which is vigorously engaged in the execution of various projects to achieve the objective of rapid socio-economic development of Himachal Pradesh.

Himachal Pradesh is bestowed with tremendous power potential in its five river basins, namely, the Satluj, the Beas, the Ravi, the Chenab and the Yamuna as depicted in Table 1.4.

TABLE 1.4
Power Potential in River Basins of Himachal Pradesh

Sl. No.	*River Basin*	*Potential*
1.	Satluj	9411 MW
2.	Beas	4597 MW
3.	Ravi	2294 MW
4.	Chenab	2748 MW
5.	Yamuna	592 MW
6.	Mini-Micro Projects	750 MW
	Total	20,392 MW

Source : *Economic Survey*, 2005, DESHP, Shimla.

It is evident from table that about 24 per cent of the total identified hydel potential of the country lies in Himachal

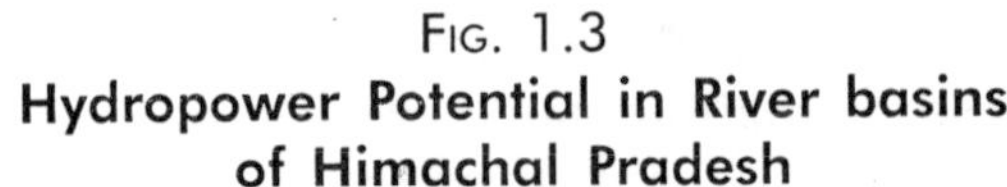

FIG. 1.3
Hydropower Potential in River basins of Himachal Pradesh

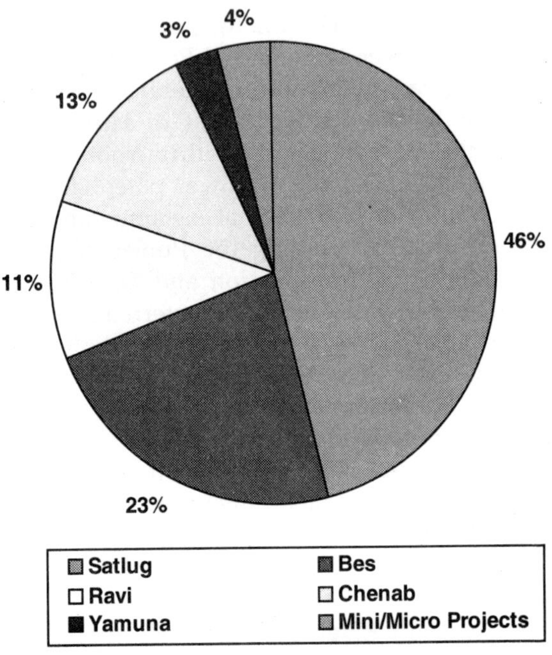

Pradesh. However, in Himachal Pradesh the hydropower development was started way back in 1912 with the commissioning of the first power plant at Chabba near Shimla to meet the power requirements of Shimla town which was then summer Capital of British India. As already mentioned, at present out of the total estimated hydropower potential of 20,392 MW, only 6060 MW has so far been exploited by various State, Central and Private Sector agencies as shown in Table 1.5. Under the multi-pronged strategy adopted by Government of Himachal Pradesh for the speedy exploitation of its vast hydel potential to mitigate the acute shortage of power being experienced in Northern region, hydroelectric projects with a total installed capacity of 8500 MW are at various stages of execution under State, Private, Central/Joint Sectors for commissioning during 10th and 11th Five-Year Plans.

1.4.1 Privatization as a Way to Induce Investment in the Sector

Capacity shortages and lack of adequate investments are considered to be among the most important challenges faced by the power sector in developing countries and privatization is advocated to facilitate the necessary investment in the energy sector. Consequently, in order to harness the hydropower potential rapidly the state is keen to encourage private sector participation in the power sector. The policy of selective privatization of power development in India was mooted by Government of India in October 1991 and Himachal Pradesh took the lead with switching over to privatization of its vast hydropower potential by advertising several projects in October 1990 followed by advertisement of more Projects in 1993, 1999, 2001 and 2003. Himachal Pradesh Government signed 28 memoranda of understanding (MOUs) with Independent Power Producers (IPPs) for various Hydroelectric Projects adding together 2868 MW between 1991 and 2004 followed by signing of Implementation Agreements (IAs) in respect of 9 Hydroelectric Projects after fulfilment of the conditions, laid down in the MOUs, by the respective companies. Out of the above 13 MOUs/I.As have subsequently been cancelled due to the failure of the IPPs to fulfil the terms and conditions stipulated in the MOUs/IAs as signed with them. In some cases even the HP Government had given a number of opportunities to the IPPs by granting extension in time limit to enable them to fulfill the terms and conditions as laid down in the MOUs/IAs. The IPPs having failed to show reasonable progress, the government was forced to cancel the MOUs/IAs signed with the IPPs and to resort to re-invitation of bids or taking up the implementation of the project in state sector. No government organization can afford to wait indefinitely for any agency to perform (Bansal, S.P.: 2005). About 13 projects are at various stages of implementation. The actual construction work was carried out only on 2 projects namely Baspa-II (300 MW) and Malana (86 MW) and the projects were commissioned in July 2001 and May, 2003 respectively contributing 386 MW by private sector till date. In addition, six mini-micro projects totaling 15MW of power generation have been commissioned during the last four years. It has been taking long time for the

TABLE 1.5

Projects under Operation in Himachal Pradesh

State Sector		*Central Sector*	
Name of Project and Year of commencement	*Installed Capacity (MW)*	*Name of Project and Year of commencement*	*Installed Capacity (MW)*
(1)	(2)	(3)	(4)
State Sector	329.5	**Central Sector**	**1038**
1. SVP Bhawa (1989)	120	1. Baira Suil (18.5.80)	198
2. Giri (1978)	60	2. Chamera Stage I (31.3.94)	540
3. Bassi (1970 and 81)	60	3. Chamera Stage II (11/03)	300
4. Ganvi (2000)	22.5	**Joint Sector**	**4050**
5. Andhra (1987)	16.95	1. Bhakra Dam (1960-68)	1200
6. Baner (1996)	12	2. Dehar (11/77-11/83)	990
7. Gaj (1996)	10.5	3. Pong Dam (01/78-2/83)	360
8. Binwa (1984)	6	4. Nathpa Jhakri (6.10.03-18.5.2004)	1500
9. Gumma (2000)	3	**Private Sector**	**401.20**
10. Nogli (1963, 1969-70 and 74)	2.5	1. Malana (5.7.2001)	86.00
11. Rongtong (1986 and 87)	2	2. Baspa-II (5/2003)	300.00
12. Sal-II (12/99)	2	3. Solang (7/2002)	1.00
13. Chaba (1912 and 18)	1.75	4. Titang (2/2002)	0.90

14. Rukti (1979 and 80)	1.5	5. Raskat (7/2001)	0.80
15. Gharola (1972)	0.05	6. Maujhi ((6/04)	4.50
16. Bhuri Singh P/House (1962)	0.45	7. Dehar (7/2004)	5.00
17. Sissu (1976)	0.10	8. Baragaon (8/2004)	3.00
18. Billing (1977)	0.20	**Others**	**241.57**
19. Shansha (1977)	0.20	1. Shanan (1931, 4/82)	110.00
20. Thirot (1995-96)	4.50	2. Yamuna Projects (HP's share)	131.57
21. Killar (1995-96)	0.30		
22. Holi (12/04)	3.00		
		Grand Total	6060

Source : Compiled from *Economic Survey*, Directorate of Economic and Statistics, Himachal Pradesh.

FIG. 1.4
Projects Under Various Sectors of Himachal Pradesh

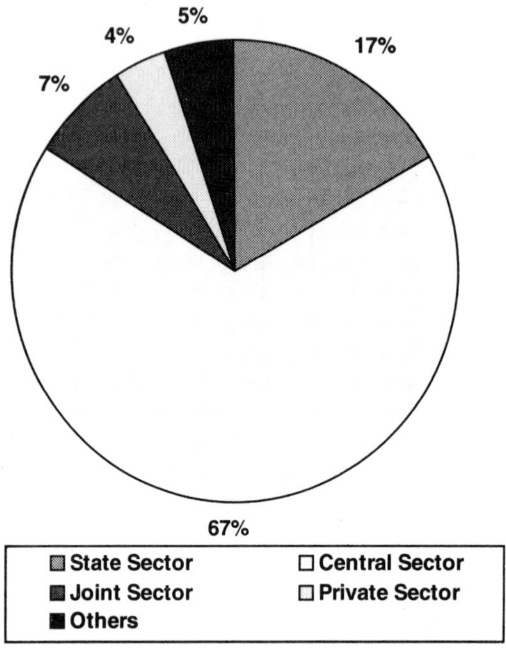

companies to obtain various statutory/non-statutory clearances, tying up for sale of power, finalization of the evacuation arrangement and achieve financial closure after tying up the necessary project loans from the Financial Institutions (FIs)/ banks.

1.4.2 Issues in Speedy Development of Hydropower Projects

For speedy development of hydropower projects, the following critical areas are required to be addressed:

Avoidance of Interference with Local Area Problems

The Government is required to ensure before award of projects that the same do not in any way interfere with the Wild Life Sanctuary, Eco-zone, Fish growth, religious sentiments of the local inhabitants, etc.

Creation of Infrastructure for Project Construction and Approach Roads

Creation of infrastructural facilities such as roads, bridges, housing for project staff and workers, telecommunication links and other supporting services consume a valuable time period of two to three years before detailed construction can be taken up. It is therefore necessary that these facilities are made available during detailed investigation stage for speedy development of the project.

Clearances

The Government is required to obtain major statutory clearances, viz. Techno-economic clearance, Environmental and Forest clearance in a time bound manner which should be transferable to the executing agency.

Identification of Land

The total land requirement for various components of the Projects needs to be identified and the basic formalities for acquisition of the same under Land Acquisition Act, 1894 should be fulfilled well before taking up the project for construction.

Rehabilitation/re-settlement

A comprehensive rehabilitation/re-settlement plan for the displaced persons in consultation with the affected families, local administration is required to be prepared in advance after conducting detailed socio-economic surveys.

Management

Modern management practices with suitable delegation of powers to enable the project staff to take immediate and quick decisions are required to be adopted. The execution of Malana Project in a record period of 33 months is a current example of comprehensive construction planning and ideal management techniques.

1.4.3 Other Benefits of Private Sector Participation

Private Sector participation in the power sector would be beneficial not in short-term basis but it has long-term prospective of sustainable economic development of the State of Himachal Pradesh. The main conditions of the agreement with private sector are as under:

- The project site is handed over for operation of the project for a period of 40 years after commercial operation.
- The project is to be transferred to the Government of Himachal Pradesh free of cost after expiry of the agreement period.
- The royalty in the shape of free power is to be given to the Government @ 12 per cent of the power generated during the entire agreement period in respect of captive use of power within the state or sale of power to the State Utility. In case of sale of power outside the State, the free power is to be given @ 12 per cent for the first 12 years after commercial operation and 18 per cent for the remaining agreement period.
- Local people should be preferred for employment in the project.

The hydroelectricity scenarios at world level, India level have been discussed comprehensively. The status of hydroelectricity in Himachal Pradesh has also been studied. Himachal Pradesh ranks high on infrastructure provision and it is the first state that achieve close to 100 per cent electricity availability for its households. Due to rapid exploitations of hydroelectric potential, economy of Himachal Pradesh is growing at rates generally above the national average with lowest poverty ratios in the country. Himachal Pradesh is among the few states with zero power deficits. The ensuing chapter is devoted to the analysis of relationship between hydroelectricity generation and economic development with special emphasis on Himachal Pradesh.

1.5 DATABASE AND METHODOLOGY

The study is based on both primary and secondary data and information collected on selected indicators in respect of persons living near the Hydropower Plants situated on the Satluj River Valley. The information and data were collected from credible and highly respected publications of Satluj Jal Vidyut Nigam Limited.

The major sources of secondary data are: Publications and unpublished official records of Himachal Pradesh State Electricity Board; Reliance Review of Energy Markets, Energy Research Group, Reliance Industries Limited; Ministry of Power, Government of India; Statistical Abstract of India-2004, Central Statistics Organization, Government of India; national Five Year Plans and State Five Year Plans (various Plan documents), Statistical Outline of Himachal Pradesh, Statistical Abstracts of Himachal Pradesh, Social Statistics of Himachal Pradesh and State Domestic Product of Himachal Pradesh (these having been published by the Department of Economics and Statistics, Government of Himachal Pradesh, Shimla).

In addition to above, data and information have also been drawn from various publications and web sites of Central Electricity Authority, Central Electricity Regulatory Commission, Ministry of Non-Conventional Energy sources and World Bank.

The most crucial task in the present study was to analyze the relationship between Hydropower Development with Economic Development. For this purpose, seven standard indicators of Hydropower Development and Economic Development (quality of life) were selected for the western Himalayan region: Himachal Pradesh. These indicators are: Net State Domestic Product at current prices, Per Capital Income at current prices, Number of Registered Factories, Number of Persons employed in Registered Factories, Generation of Power, Consumption of Power and Consumption of Power in Industrial sector.

The degree of relationship may be measured and represented by the Coefficient of Correlation. Therefore, in the present study method of Karl Pearson's Coefficient of Correlation has been applied to ascertain the relationship between power sector and economic development.

The method of *Karl Pearson's Coefficient of Correlation* is given on the next page.

$$r = \frac{N\Sigma XY - (\Sigma X)(\Sigma Y)}{\sqrt{N\Sigma X^2 - (\Sigma X)^2 N\Sigma Y^2 - (\Sigma Y)^2}}$$

where SX = Sum of the X Scores
SY = Sum of the Y Scores
SX^2 = Sum of squared X Scores
SY^2 = Sum of squared Y Scores
SXY= Sum of the products of paired X and Y scores
N = Number of paired scores.

The value of coefficient of correlation for each indicator has also been statistically tested so that relevance of relationship can be justified.

An effort has also been made to determine the role of the Hydroelectric Power Projects of SJVNL on Socio-Economic Development of Project Affected Area/families. For this purpose, measurement of Level and Status of Development Index of SJVN Hydro Power Projects Affected Area *vis-a-vis* districts of Himachal Pradesh was essential. In the present study, most appropriate method which is being used by *United Nations Institute of Social Research (UNISR: 1991) for measuring social development* has been applied. The index is derived as follows:

$$\text{Deprivation Score} = \frac{\text{Maximum Value} - \text{Actual Value}}{\text{Maximum Value} - \text{Minimum Value}} \quad \ldots (1)$$

The deprivation score ranges between 1, denoting maximum deprivation, to 0 representing least deprivation. Subtracting the deprivation score from 1, the development index is worked out as:

$$\text{Development Index} = 1 - \text{Deprivation Score} \quad \ldots (2)$$

The development index ranges between 1, denoting maximum development, to 0 representing no development. It measures the relative development level of various regions in relation to the most developed one. This measures the extent to which a particular region is lagging behind as compared to the one at the top. The UNISR, however, makes use of a simpler formula. The development score is calculated directly by ascertaining the position of each region *vis-à-vis* the most backward ones. The results, however, are virtually the same by both the methods.

1.6 MAIN HIGHLIGHTS

- The human beings since their inception possesses remarkable quality of thinking rationally behave properly and adjust accordingly to their environment due to their highly developed brain. For this reason, they fall on the top hierarchies of the primates, who really can experience bliss from their surrounding environment. They utilize their cognition with the help of emergent brain activities in exploring their environment. Today technology has become so advanced that it has helped human beings in every gamut of their life. The international community set itself the Millennium Development goal to make available the benefits of new technologies.
- Hydroelectricity is a renewable source. It uses the energy of running water, without reducing its quantity, to produce electricity. Therefore, all hydroelectric developments, of small or large size, whether run of the river or of accumulated storage, fit the concept of renewable energy.
- Hydroelectric power plants with accumulation reservoirs offer incomparable operational flexibility, since they can immediately respond to fluctuations in the demand for electricity. The flexibility and storage capacity of hydroelectric power plants make them more efficient and economical in supporting the use of intermittent sources of renewable energy, such as solar energy or Aeolian energy.
- River water is a domestic resource which, contrary to fuel or natural gas, is not subject to market fluctuations in addition to this, is the only large renewable source of electricity and its cost-benefit ratio, efficiency, flexibility and reliability assist in optimizing the use of thermal power plants.

- Hydroelectric power plant reservoirs collect rainwater, which can then be used for consumption or for irrigation. In storing water, they protect the water tables against depletion and reduce our vulnerability to floods and droughts.
- The operation of electricity systems depends on rapid and flexible generation sources to meet peak demands, maintain the system voltage levels, and quickly re-establish supply after a blackout. Energy generated by hydroelectric installations can be injected into the electricity system faster than that of any other energy source. The capacity of hydroelectric systems to reach maximum production from zero in a rapid and foreseeable manner makes them exceptionally appropriate for addressing alterations in the consumption and providing ancillary services to the electricity system, thus maintaining the balance between electricity supply and demand.
- The hydroelectric life cycle produces very small amounts of greenhouse gases. Hydroelectric power plants don't release pollutants into air. They very frequently substitute the generation from fossil fuels, thus reducing acid rain and smog. In addition to this, hydroelectric developments don't generate toxic by-products.
- The hydroelectricity is the largest renewable resource of energy that according to UNDP has gross theoretical potential at about 40,500 Tera Watt Hours, technical potential at about 14,320 TWh and economic potential at about 8100 TWh. The installed capacity of the world currently stands at 694 GW (6,94,000 MW) and by 2000 the generation of hydroelectricity was 2675 Billion Kwh or close to 20% of the total electricity generation. Canada, Brazil, US, China and Russia were the top five produces while India stands at the ranking of 8th number. Total global installed capacity by 2000 was 3262 GW out of which 2175 GW fossil fuels account 67%; 694 hydroelectric or 21%; 358 GW Nuclear or 11% and other or 1%. By 2020 the electric

Power Generation is expected to be about 25,500 TWh. The installed capacity of top five country include US, China, Japan, Russia, Canada and India ranked at 8th number in the world.

- India is becoming self-sufficient and self-reliant by inclining its focus toward the application of science and technology and diverting her attention toward exploiting various types of energy in general and hydroelectricity in particular. Various studies are under progress for the betterment and sustainability of the Nation. One such study was conducted by Water and Power Commission in India on hydroelectricity during 1953-59 that assessed its potential here as 42,100 MW at 60% load factor corresponding to 221 billion unit's energy annually. The study was re-assessed by Central Electricity Authority in 1987 that placed hydro potential of the country at 84,044 MW at 60% load factor. Approximately 845 hydroelectric schemes have been identified in the various basins and expected to yield approximately 442 billion units of electricity.
- Power generation in Himachal Pradesh can become a significant catalysts and driving force for economic development. In course of hydrological, topographical and geological investigation this state has been projected about 21,244 MW of hydel power that can be generated here. Out of the total hydel potential, only 6060 MW has been harnessed so far in Himachal Pradesh out of which only 329.50 MW is under the tight control of Himachal Pradesh and bulk of potential has been exploited by the Central Government and other Agencies. The Satluj Jal Vidyut Nigam Limited was incorporated in May 24, 1988 as a joint venture of the Government of India and Himachal Pradesh. It has been playing very important role in planning, investigating, organizing, executing, operating and maintaining hydro-electric power projects on the river Satluj basin in this state. This organization has bagged number of awards and recently in January 2008 it received Golden Peacock

award for the judicious settlement of dumping waste materials of Rampur hydroelectric project in Himachal Pradesh (Divya Himachal, January, 2008). This Nigam has already completed 1500 MW Nathpa Jhakri Project and is presently executing Rampur Hydel Project (412 MW) while Luhri hydel projects (700 MW) and Khab project (1020 MW) will likely to be executed very soon in near future.

Notes and References

1. Electricity is used everywhere and there is no place to left out and its applications in home, factories, and transportation are quite commendable. In home it is used for cooking food, heating, lightening, and running radio, T.V. sets, refrigerators, and microwave oven, etc. In factory it is used for pouring heavy scale machinery and in transport for giving power to the buses and railways.
2. The worldwide theoretical, technical and economic potential of hydropower is according the study of UNDP and other studies conducted for the purpose of estimation of hydropower potential in the world.
3. In the EIA's reference case scenario, worldwide consumption of renewable energy is projected to increase by 53% between 1999-2020, compared to 92% increase for natural gas and 58% for oil during the same time frame, with large-scale hydroelectricity installations providing for bulk of the growth in renewable energy use in the developing world (Reliance Review of Energy Market, 2002, p. 193).
4. In some of the future long-term energy scenarios, Biomass is expected to contribute substantially to the global mix by the year 2100, anywhere between 2500 Million Tons of Oil Equivalent (MTOE) to 7500 MTOE. Biomass energy technology is also rapidly advancing. Moreover direct combustion, techniques for gasification, fermentation and anaerobic digestion are all increasing the potential of Biomass as a substantial energy source.
5. The technical potential of onshore wind energy is very large between 20,000 to 50,000 TWh per year against the current total annual world electricity consumption of about 15,000 TWh. The economic potential depends upon factors like average wind speed, statistical wind speed distribution, turbulence intensities and the costs of wind turbine systems. Because the energy of the wind is proportional to the third power of the wind speed, the economic calculations are very sensitive to the local average annual wind speed.
6. The UNDP study expects an installed capacity of about 20,000 MW by 2002 and 30,000 MW by 2004. Thereafter, based on a 15 per cent cost reduction, and later 12 per cent and 10 per cent for each doubling of the accumulated number of installations—the study predicts a penetration of 10 per cent of the world's electricity needs by winder power around the

year 2025 (For further detail, please see Reliance Review of Energy Market, 2002).

7. Photovoltaic solar energy conversion is the direct conversion of sunlight into electricity. An essential component of these systems are the solar cell, in which the photovoltaic effect, i.e. the generation of free electrons using the energy of light particles take place. These electrons are then used to generate electricity.
8. In 1982, a separate Department of Non-Conventional Energy Sources was created in the Ministry of Energy and entrusted with the charge of promoting non-conventional energy sources. In 1992, this was upgraded into a separate Ministry of Non-Conventional Energy Sources (MNES) to develop various areas of renewable energy. The Ministry is broadly organized into six groups, viz. Rural Energy, Solar Energy, Power from Renewables, and Energy from Urban and Industrial wastes, New Technologies and Administration and Coordination.
9. The NTPC Ltd. (formerly National Thermal Power Corporation Ltd.) was incorporated in November 1975 with the objective of planning, promoting and organizing integrated development of thermal power in the country. The Company has now been renamed as NTPC Ltd. In line with the changes taking place in the business portfolio of the company transforming the company from a thermal power generator to an integrated Power Company with presence across entire energy value chain. NTPC is a schedule 'A' Navratna company having a total approved investment of Rs. 91619.92 crore, as on 31 March 2006. The Corporation has under operation/implementation coal-based super thermal power projects at 16 locations and combined cycle gas power projects at 7 locations, including 705 MW Badarpur Thermal Power Station (BTPS) in Delhi which NTPC had been managing on behalf of Government of India has since been taken over by NTPC w.e.f. 01 June 2006. The commissioned capacity of NTPC owned stations as on 30 June 2006 is 24,640 MW. In addition, it has acquired 314 MW of captive power plants of Steel Authority of India Limited (SAIL) through formation of joint venture companies with it. NTPC is at present implementing 10 power projects with a capacity of 9470 MW. Corporate Plan adopted by NTPC envisages it to become a 46,000 MW plus company by the year 2012 and 66,000 MW plus company by 2017. During the year 2005-06 an all time high generation of over 170,880 MUs was achieved registering an increase of 7.4 per cent over the previous year's generation of 159,110 MUs. NTPC coal based stations achieved ever-highest PLF of 87.54 per cent since inception during the year.
10. As on 31 March 2006, PGCIL is operating about 55,120 ckt kms of transmission lines consisting of 563 ckt. kms. of 800 KV, 4,368 ckt. kms, of HVDC system, 40,179 ckt. kms. of 400 KV, 7,734 ckt, kms. of 220 KV and 2, 241 ckt. kms., of 132 KV and 37 ckt. kms. of 66 KV lines along with 93 sub-stations with about 54, 380 MVA transformation capacity. The transmission system availability is maintained consistently over 99 per cent by deploying best operation and maintenance practices at par with international utilities.
11. Bhakra-Nangal Project proposal was originated in 1908 by Sir Lious Dane the then governor of Punjab. He preferred two sites one in Suni (District

Shimla) and other in Bhakra (Panjab) and finally selected the latter one. It was the ambitious project in rugged terrain of erstwhile Bilaspur principality and the decision of height of the dam was estimated ass 200 feet. Up to 1938-39 the plan remains the pending due to heavy expenditure. In the same year, severe drought in Rohtak and Hissar resulted great loss of life and property including livestock. The Bhakra dam scheme was initiated, the reservoir was planned to supply water to dry district and also to provide energy in order to accelerate the process of economic development in the fundamentally backward economy of Northern India. But World War-II hindered the execution of the scheme. In 1944 Dr. J.L. Savage of USA re-visited the dam site and he allowed execution of the project. Consequently, a famous dam builder, Harvey Slocum engineered the gigantic project. The reservoir was decided at Bilaspur district of Himachal Pradesh, presently it is known as Govindsagar Lake. With the construction of this project a large number of families were displaced for the cause of National development. The famous economist Raj (1960) had made an attempt to analyze social cost of this project by taking into account of socio-economic indicators only; however, the crucial psychological cost of thousands of displaced people was not studied.

12. However, it is important to recognize that small hydro capacity spans three categories: micro (less that 100 kW), mini (100 kW-2,000 kW) and small (2,000 kW-25,000 kW). At the low end of the range, simplicity of design and control is usually essential for economic viability while at the high end of the range; the magnitude of the investment is likely to warrant fairly sophisticated protection and control devices. Seasonal water flow variations can be overcome to some extent by incorporating a storage reservoir in the hydro scheme and adapting water release through the power station to the requirement of the power market. The installed, capacity can then be more closely tailored to the electricity demand. For run-of-river projects where no significant water storage exists, the firm capacity will be limited by low-flow water conditions and will only be a small fraction of the installed capacity. In such cases the value of electricity generated in equivalent to the displacement energy value only; nevertheless, it is usually not economically justifiable to create a storage reservoir solely to serve a small hydro plant.
13. The basic objectives of the Rural Electrification Corporation are: (1) To promote and finance projects aimed at integrated system Improvement, power generation, promotion of decentralized and non-conventional energy sources, energy conservation, renovation and maintenance, power distribution with focus on pump-set energisation, implementation of Government of India scheme for electrification of all villages and all households by 2009 and other related works in rural and urban areas; (2) To expand and diversify into other related areas and activities like financing of decentralized power generation projects, use of new and renewable energy sources, consultancy services, transmission, sub-transmission and distribution systems, renovation, maintenance and modernization, etc., for optimization of reliability of power supply to rural and urban areas, including remote, hill, desert, tribal, riverine and other difficult/remote areas; (3) To mobilize funds from different sources

including raising of funds from domestic and international agencies and sanction loans to the State Electricity Boards, State Governments, Power utilities, Rural Electric Cooperatives, Non-Government Organizations (NGOs) and private developers; (4) To optimize the rate of economic and financial returns for its operations while fulfilling the corporate goals viz. (i) laying of power infrastructure; (ii) power load development; (iii) rapid socio-economic development of rural and urban areas, and (iv) technology upgradation; (5) To ensure client satisfaction and safeguard customers' interests through mutual trust and self-respect within the organization as well as with business partners by effecting continuous improvement in operations and providing the requisite services; and (6) To assist State Electricity Boards/Power Utilities/State Governments, Rural Electric Cooperatives and other loanees by providing technical guidance, consultancy services and training facilities for formulation of economically/financially viable schemes and for accelerating the growth of rural and urban areas. Net profit of REC during 2004-05 was Rs. 801 crore, net worth of the company is Rs. 3779 crore and loans disbursed during the year 2004-05 by REC was Rs. 7885 crore.

14. The NPTI also established high fidelity, Real-Time, Full-Scope 500 MW and 210 MW Fossil Fuel Fired Power Plant Training Simulators imparting training to nearly 8,000 engineers and operators across the country. Also a 430 MW CCGT Replica Simulator has been commissioned at the Corporate Office, Faridabad. A GIS Resource Centre has also been established for Training purposes at Faridabad. Over 40 self-paced, menu-driven, cost-effective multimedia computers-based training packages have been developed and marketed by NPTI. In its attempts to weave formal education with Industry-oriented inputs, NPTI is conducting: (a) Two year MBA in Power Management, (b) Four-year Degree Course in B.Tech/B.E. (Power), (c) Post Graduate Diploma in Thermal Power Plant Engineering. These AICTE approved courses have an overwhelming response and the trained manpower is being recruited by various Public/Private Sector Organizations through Campus interviews. Besides these NPTI is also conducting a Post Diploma Course in Thermal power Plant Engineering.
15. The Institute, with its existence for over four decades has built sophisticated facilities, both in the areas of research and testing. The important facilities include 2500 MVA Short Circuit Testing with Synthetic Testing Facility at Bangalore, Ultra High Voltage Research Laboratory at Hyderabad; Short Circuit Testing Facility at Bhopal; Thermal Research Centre at Koradi, Nagpur; and Energy Research Centre at Thiruvananthapuram to cater to the R&D and testing needs of the power sector. A state-of-the–art test facility for seismic qualification of power equipment and Real Time Digital Simulation (RTDS) facility have been set-up and commissioned. The CPRI's laboratories are accredited under National Accreditation Board for Testing and Calibration of Laboratories (NABL), which is the national body for accreditation of laboratories as per ISO/IEC 17025 norms. CPRI's low and medium voltage laboratories are accredited by ASTA BEAB, UK. CPRI has been given the 'observer status' in the group of Short Circuit

Testing Liaison (STL) of Europe and is likely to be a full member of this prestigious club. CPRI laboratories are approved for testing for certain products like communication cables, L.T. Capacitors for motors, etc., by Underwriters Laboratories (UL) and Canadian Standard Association (CSA). CPRI is also CBTI, under IEC-EFCB Scheme. CPRI's R&D and Consultancy activities have been granted ISO 9001:2000 Certificates.

2

Hydroelectricity and Economic Development

There is no denying of the fact that energy is an essential input for economic development and as such it facilitates in improving the quality of life of the general population. Development of conventional forms of energy for meeting the growing energy needs of society at a reasonable cost is responsibility of the Government, Development and promotion of non-conventional/alternate/new and renewable sources of energy such as solar, wind and bio-energy, etc., are also getting sustained attention. Nuclear energy development is being geared up to contribute significantly to the overall energy availability in the country (India: 2006). In view of the above facts, the present chapter has four vital objectives. First part of this chapter is devoted to emphasize the role of hydroelectricity in the process of economic development. Second section discusses the role of hydroelectricity in improving quality of life. Thirdly, it explains hydroelectricity and rural development. Fourthly, it elaborately provides the plan efforts for the development of power sector in India as well as in Himachal Pradesh.

2.1 HYDROELECTRIC POWER AND ECONOMIC DEVELOPMENT

Electricity is a vital infrastructure for economic development. There is a close correlation between the per capita GNP of a country and the per capita electricity it consumes. It is a well established fact that electricity and economic growth go hand in hand. In fact, some economists regard electricity as the fourth factor of production, in addition to the traditional listing of land, labour and capital. Some of the major areas where electricity has substantial impact on development are mining, irrigation, transportation, production of consumer goods, cement, steel, domestic appliances, recycling of waste materials into useful ones, etc.

Development of hydroelectric power, interpreted broadly to indicate increased provision and use of energy services, is an integral part of enhanced economic development. Advanced industrialized societies use more electric power per unit of economic output and far more electricity per capita than poorer societies, especially those still in a pre-industrial state. Electricity use per unit of output does seem to decline over time in the more advanced stages of industrialization, reflecting the adoption of progressively more proficient technologies for energy production and utilization as well as change in the composition of economic activity. Electricity intensity in developing countries perhaps peaks faster and at a lower level along the development path than was the case during the industrialization of the developed world. But even with trends toward greater energy efficiency and other dampening factors, total energy use and energy use per capita continue to grow in the advanced industrialized countries, and even more rapid growth can be expected in the developing countries as their incomes advance. Development involves a number of other steps besides those associated with energy, notably including the evolution of education and labour markets, financial institutions to support capital investment, modernization of agriculture, and provision of infrastructure for water, sanitation, and communications. This is not just an academic question; energy development competes with other development opportunities in the allocation of scarce opportunities for policy and institutional reform.

Accelerating economic growth and achieving higher standards of living depend upon the availability of adequate and reliable power at an affordable price. Unlike other commodities, electricity cannot be stored for future use. In other words, its generation and consumption have to be simultaneous and instantaneous. Furthermore, electricity, a major source of energy, is regarded as an important factor for bringing about radical change in the socio-economic life of a community. Because of multifarious uses of electricity, such as for lightning and as a source of motive power, its introduction does not merely facilitate provision of better amenities but augments productive capacity in different sectors of the economy through its wide range of applications. The availability of electric power affects industries in two principal ways. Firstly, the traditionally operated industries gradually change over to the use of electric power, and secondly, new industrial units operated by power come up. Use of electric power, therefore, strengthens and expands both small and large industrial sectors and makes them more efficient and productive.

Similarly, rural electrification has a significant and positive impact on agricultural production, thereby creating additional processible resources. Rural Electrification Programme is not to be seen as a drive to light the house of the people. Two related aspects should be emphasized in this context. One is the problem of finding alternative sources of energy as hitherto complete dependence on forests cannot continue for long for obvious reasons. Therefore, the rural electrification scheme has to be viewed as a part of the programme for a search of alternative sources of energy and its appropriate utilization. The second aspect is related to the direct benefits of electricity in underdeveloped rural areas.

Therefore, power development has a significant role in promoting socio-economic development by creating a base on which a higher level of economic activity can be carried out. Energy is like blood in human body. Deficiency of blood in human body reduces not only physical power but also mental ability. In the similar way adequate energy supply is a *sine-qua-non* for rapid and sustainable economic development. Social and economic aspects of population, especially those who reside near the project site, are one of the most important considerations for any development project. The construction phase of a project generally has a pronounced impact on the socio-economic aspects on the local population for the reasons that a section of population faces displacement from their land which is acquired for the construction of project and the construction activities brings in a sea change in the local environmental settings due to large scale influx of population altering the traditional economic activities, etc. The construction sites of hydropower projects have shown inward migration of large number of labour force including drastic change in the socio-economic fabrics of the local population (Tiwari, 2007).[1]

Along with economic growth, electricity consumption increases as power has become an integral part of improved and modernized infrastructure for production as well as consumption. The per capital consumption of electricity, therefore, reflects the level of such improvement and modernization, in short, of development. States that have provided a higher percentage of households with electricity connections[2] have given their citizens better access to growth prospects. In order to substantiate the relationship between availability of electricity and economic development, this hypothesis has already tested by Tiwari (2007) on Himachal Pradesh where there is vast potential for hydropower generation due to its geographical settings. This micro-level analysis on Himachal Pradesh eventually leads to generalize the hypothesis of symbiotic relationship between energy generation/consumption and economic progress at the macro-level.

2.1.1 Relationship between Availability of Hydro-electricity and Economic Development: Evidence from Himachal Pradesh

The state of Himachal Pradesh has been averaging an annual economic growth rate of around 7 per cent since 1994-95, a commendable achievement, generally, at rates above the national gross domestic product growth. With a per capita income of Rs. 30,138 in 2005-06, it ranks seventh among the states, with a rate of growth of per capita income higher than the all-India trend. Himachal Pradesh has one of the lowest levels of poverty in the country with a mere 7.6 per cent of its population living below poverty line in 1998-99, compared to the all-India average of 26.1 per cent. The economy is largely dependent on its service sector, which contributes 39.6 per cent to total income, though the manufacturing sector has also been vibrant as it makes up 34.5 per cent of the state income. As with other states, over time its economy has shifted from being more dependent on agriculture sector to focusing on industry and the service sector.

Himachal Pradesh ranks high on infrastructure provision in the country as it is the first state that has achieved close to 100 per cent electricity connections for its households. Given the topography of the state, the fact that 90 per cent of its population resides in rural areas and the fact that these 20,000 odd villages are scattered, this is a remarkable achievement. The state is among the few states with zero power deficits. It has a huge potential of hydropower generation at 2,03,086 MW which is nearly 24 per cent of the total potential in the country.

In this section, an effort has been made to analyse the relationship between hydropower potential and economic development in Himachal Pradesh on a time-series data as presented in Table 2.1. Correlation is a measure of relationship between two variables. In the present case method of partial correlation has been applied. In partial correlation, we are interested in knowing the extent of specific relationship between two variables, factoring out the effect of one or more other variables. The value of correlation coefficient and r-value for two-tailed test of significance are presented in Table 2.2.

The correlations of following pairs are given in Table 2.2:

- Net State Domestic Product at current prices and Per Capita income at current prices.
- Net State Domestic Product at current prices and number of registered factories.
- Net State Domestic Product at current prices and number of persons employed in registered factories.

Table 2.1
Variables of Hydropower Development and Economic Development

Year	NSDP (at current price)	Per capita income (Rs.) (at current price)	No. of Regd. Factories	Employment in Industrial Sector	Hydropower Generation (in lakh kWh)	Electricity Consumption (in lakh kWh)	Electricity consumption in industrial sector (in lakh kWh)
(1)	(2)	(3)	(4)	(5)	(6)	(7)	(8)
1971	223	651	174	11214	528.41	1119.64	124
1981	722.82	1704	549	15700	2450.66	2859.62	1076
1991	2521.47	4901	1308	34568	12624	10087	5996
1995	5192.46	9451	1495	43783	11317	13396.8	7973
1996	5930.24	10607	1556	48970	12854.2	15980	9670
1997	6802.87	11960	1657	60525	12520	17571	10599
1998	7806	13488	1694	64544	13060	19465	11825
1999	9507.46	16144	1780	69215	14845	20834	12493
2000	10881.5	18160	1972	76336	12013	21817	12594
2001	12022.6	19925	2010	77620	11533	22058	12772
2002	13336.6	21570	2097	80114	11495.01	23318	13248
2003	14262.4	22671	2079	80608	12779.29	25165	14526

Source : Statistical Outline of Himachal Pradesh (various issues), DESHP, Shimla.

TABLE 2.2
Inter-correlation Matrix

Indicators	*SDP*	*PCI*	*No. of Factories*	*Employment in Factories*	*Hydropower Generation*	*Electricity Consumption*	*Elec. Consumption in Industrial Sector*
(1)	*(2)*	*(3)*	*(4)*	*(5)*	*(6)*	*(7)*	*(8)*
SDP	1.00	.999**	.915**	.972**	.653*	.960**	.939**
PCI		1.00	.931**	.981**	.683*	.971**	.953**
No. of Factories			1.00	.965**	.879**	.978**	.979**
Employment in Factories				1.00	.775**	.992**	.985**
Hydropower Generation					1.00	.827**	.858**
Electricity Consumption						1.00	.998**
Elec. Consumption in Industrial Sector							1.00

* Correlation is significant at the 0.05 level (2-tailed).

** Correlation is significant at the 0.01 level (2-tailed).

FIG. 2.1

Trend of Hydropower and Economic Development in H.P.

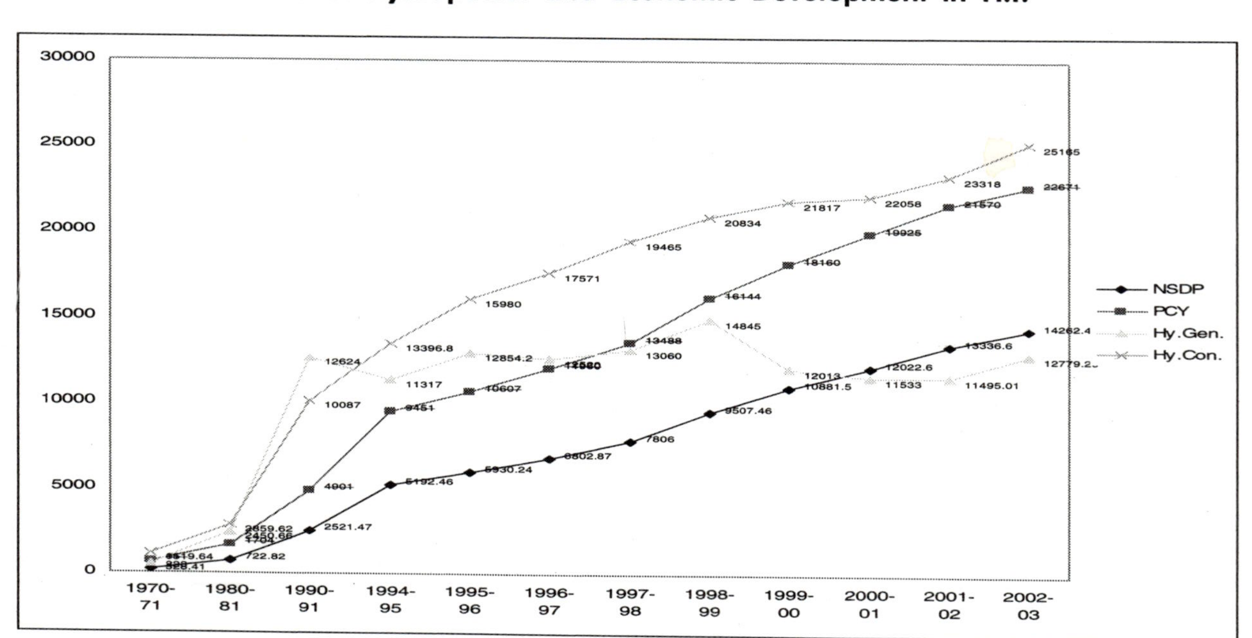

Based on Table 2.1.

- Net State Domestic Product at current prices and hydropower generation.
- Net State Domestic Product at current prices and electricity consumption.
- Net State Domestic Product at current prices and consumption of electricity in industrial sector.
- Per Capita Income at current prices and number of registered factories.
- Per Capita Income at current prices and employment in registered factories.
- Per Capita Income at current prices and hydropower generation.
- Per Capita Income at current prices and hydropower consumption.
- Per Capita Income at current prices and electricity consumption in industrial sector.
- Number of registered factories and number of persons employed in registered factories.
- Number of registered factories and hydropower generation.
- Number of registered factories and hydropower consumption.
- Number of registered factories and consumption of electricity in industrial sector.
- Employment in registered factories and hydropower generation.
- Employment in registered factories and hydropower consumption.
- Employment in registered factories and electricity consumption in industrial sector.
- Hydropower generation and hydropower consumption.
- Hydropower generation and consumption of electricity in industrial sector.
- General Hydropower consumption and consumption of electricity in industrial sector.

It is noticeable from Table 2.2 that relationship between power sector (In the present context, three indicators are representing power sector, viz. Power generation, Power

Consumption in general and consumption of power in industrial sector) and economic development (the remaining indicators generally pertain to economic development) was positive and significant, which obviously proves the hypothesis of strong association among per capita income, per capita consumption of electricity and industrialization of Himachal Pradesh. There is no denying the fact that the way the Government of Himachal Pradesh has embarked upon an accelerated Power Development Programme, the State is speedily moving towards a "Power State" of the Country and it is the first state to achieve close to 100 per cent electricity availability for its households. Himachal Pradesh is a state that has shown how outcomes, and not outlays, are crucial for growth and development. As a model state, it has its own levels of achievements to surpass in the future as it strives to stay on a high growth trajectory.

The findings of the studies of Bajpai (1985), Tiwari (1988) and Ahluwalia (1985) brought out the role of industrial infrastructure, particularly power and coal, in the growth of industrial production. Similarly, the study of Tiwari (2007) also points towards a positive relationship between hydroelectricity and economic development in Himachal Pradesh. Further, he studied the role of Nathpa-Jhakri Electric Power Project in the Economic Development of Himachal Pradesh. It was concluded that the status of socio-economic development of Project Affected people was better not only in comparison to the people of Shimla and Kinnaur districts but in respect of the State average also. In the matter of level of development, (where NJPC Project areas was taken as separate unit just like districts of Himachal Pradesh for the sake of meaningful analysis of level of development) the NJPC Project Affected People was placed in high level of development in respect of sex ratio, literacy percentage, female literacy percentage, literacy percentage among SC/ST population, female literacy percentage among SC/ST female population, percentage of main workers to total population, percentage of female main workers to total female population and percentage of families placed above the poverty line.

2.2 HYDROELECTRICITY AND QUALITY OF LIFE

Quality of Life refers to the degree of excellence in one's life at any period of time that contributes to satisfaction and happiness of the person and benefits the society. It involves two factors named as satisfactory conditions and satisfying conditions. The satisfactory conditions include factors like group cohesiveness, sharing of each other's experience, helping attitudes, understanding and sharing each other's problem and absence of any physical and mental illness, etc. The satisfying conditions include factors such as sense of belongingness, presence of positive attitudes; subjective feeling to the physical well-being and absence of unhealthy experiences, etc. (Verma and Asthana, 1986). The advancement of technology is directly associated with the better quality of life and is becoming a significant catalysts in the improving economy and wellbeing of the people. In India, the Center, State, Public, Private, Profit and non-profit-making organizations have been adopting the implications of scientific technology for its sustainable development by various means.

The quality of life means improving the conditions in which people live and providing them the environment that enables them to lead life as they would like to live it are universally accepted societal goals. The very objective of development everywhere has been to promote these goals, although those responsible for development like the planners or aid-donors may define the ends that they are pursuing for the people only in rather minimalist terms like raising the standard of living of the people and may not go to the core of what the people and the society would be really looking for as the ultimate level of fulfilment for them.

There is no denying the fact that human beings seek fulfilment through improvements in their quality of life. The quality of life is denoted by human well-being or well-lived life (Dasgupta, 2001), a state of thriving "good human living", or a "good life", etc. (Nussbaum & Sen, 2002). Analysts have gone into the question of what constitutes such a life and what promotes it. Interestingly the definition of quality of life, i.e. (human well-being) its components and determinants, and the methods of measuring it have generated a great deal of debate.

Among the various facets of analysis of quality of life, the list contentious seems to be the components and the determinants of human well-being. According to Dasgupta a well lived life has some basic features, such as personal growth (promoted by self-respect, honesty, cheritability, displaying and receiving of affection, among other features), a successful family life, warm friendship, a satisfying job, an occasional trip to see other places and culture, a meaningful social life, satisfaction of commodity requirements and enjoyment of human rights. Amartya Sen refers to some "valuable functioning" in terms of which quality of life can be assessed. Some of these are very elementary and basic but strongly valued by all, such as being adequately nourished and well sheltered, being in good health, etc. Others are more complex like being happy, achieving self-respect, being socially integrated, appearing in public life without shame, and so on (Sen, 2002). He argues that in very broad terms quality of life and well-being of a person are promoted by anything that enhances freedom to live the way one would like, thus living freely and availability of freedom of choosing between alternatives.

Apart from the above, Jean Dreze and Amartya Sen (2002) have referred to social equity, removal of deprivation of the poor people, integrity of the physical environment, delivery of public services, removal of gender biases and disparities, etc. as essential for enhancement for quality of life of the people. It is perhaps clear by now that quality of life is synonymous with human well-being and happy and good life.

It is not uncommon to notice that quality of life, human well-being, level of welfare, and standard of living are terms that are usually used interchangeably. For instance, Sweden and other Scandinavian countries have been conducting standard of living surveys and in that context the foregoing terms have been used without bringing any distinction among them. However, the standard of living seen by Amartya Sen to be only a component of wider concept of human well-being. It in fact refers to those influences on well-being that come from the nature of a person on life. Robert Eriksson gives a definition of level of living, based on European tradition which perceives it as ". . . individual's command over resources in the form of money, possession, knowledge, mental and physical energies, social relations, security, and so on, through which the individual can control and consciously direct his living conditions. However, individual's overall well-being and quality of life are determined by a wider spectrum of processes. These are not only influenced by standard of living which focuses on individual's on life and circumstances and his access to economic resources, but also by what Amartya Sen calls "success in the pursuit of all the objectives that he has reason to promote". Thus, well-being can also be "other regarding". People too derive satisfaction from helping other. It is therefore clear that quality of life is influenced and determined by a wide array of factors (Sharma, 2005). While gauging the quality of life of a person or a community, the list of such factors would be quite formidable since one shall have to look at all those influences and phenomenon that add to the wellness and welfare of the people.

Quality of life is, in the ultimate analysis, a subjective phenomenon because it has to do with people's utility, feeling of wellness, and satisfactions. But in order to ordinarily evaluate the improvements or deterioration in it objective criteria such as the literacy rate, income level and life expectancy at birth have been used. In the evaluation of subjective phenomenon with the help of objective criteria a serious dilemma arises. These criteria for evaluation of human well-being are designed by various categories of experts on the basis of their judgment of what should best reflect and measure changes in the people wellness. The question is do these criteria exactly and faithfully bring out the inner feelings of a person or a community whose well-being is sought to be evaluated? Alternatively, are the people themselves not the best judges of how they feel about whether a change of policy, circumstances, or availability of additional resources have added to their well-being, and if it has, then how much? In the context of this dilemma in the measurement of phenomenon like a standard of lying, well-being and quality of life with the help of objective indicators designed by the experts on the one hand and alternatively trying to fathom the actual feelings of the people themselves on the other, Erik Allardt (2002) poses the question about common indicator of well-being viz. dwelling space. "In measuring housing standard for instance, should one rely on objective measures of the space available and the number of household's appliances in the family or should one ask whether or not the respondents are satisfied with their housing conditions?" This is a real issue in the evaluation and measurement in the change in the quality of life of the people. However, one has to assume that the objective indicators well roughly help in gauging the direction of change, whether improvement or deterioration, in the quality of life.

2.2.1 Qualilty of Life in the Himalayan Region

In a study, L.R. Sharma (2005) has made an attempt to measure the quality of life in the Himalayan states with the help of 23 indicators. The statistical data on quality of life in each of state of the Himalayan region is presented in Appendix-II for perusal. These indicators are divided into six broad categories, viz. those relating to income, consumptions and distributive justice, literacy and education, health and demographic features, those which particularly impinges on the quality of life of women who constitutes nearly 50% of the population, those relating to the basic amenities available to the household and finally, the environmental factors. It was assumed that indices of quality of life taken together will indicate the broad level of well-being of the people living in the Himalayan regions. He concluded that the states of the region which score the first and the last rankings in respect of most of the indicators have been identified. Those which score the highest number of first ranking would be considered to manifest a relatively better quality of life of its/their people, and those remaining laggards in most cases would indicate a rather low level of well-being of people. On this basis, it is noted that the states of Himachal Pradesh and Mizoram in the Himalayan region seem to enjoy a relatively better quality of life, but on the other hand, the state of Jammu & Kashmir lags behind the rest of the region in most dimensions of well-being considered in the study. He, however, noted that quality of life is a multi-dimensional concept; as such consideration of a few quantifiable phenomena and objective indicators would not help in drawing a firm conclusion on overall level of well-being of the people of a region.

According to Amartya Sen (1987), the standard of living of a person essentially refers to what that person has the opportunity to be or do (as opposed to his or her possessions, or utility, or happiness). At a general level, the standard of living of an individual can be seen to depend on his or her personal characteristics, as well as on his or her entitlement to the commodities (marketable or non-marketable) that make the relevant activities possible.

While the demand for restoration of the pre-project standard of living has been general, it has been particular prominent in World Bank document, including its operational guidelines. The World Bank, however, adopts a rather narrow interpretation of the standard of living focusing primarily on income restoration (and to some extent also on 'livelihood replacement'), it is argued, tops; the priority list of displaced persons. The focus of income restoration is also seen as a rational response to the widely observed impoverishment of project affected person. According to World Bank findings when people are compensated with land and jobs (rather than just cash), and given institutional assistance, successful income restoration is typically achieved after a transition period. This is especially to when the displaced are given a share in the immediate benefits created by the project itself.

The study conducted by Tiwari (2007) on Nathpa Jhakri Hydel Project Affected families also highlights that the status of socio-economic development of project affected families in comparison to those in Shimla (state capital) and Kinnaur districts as well as the state average was comparably better. In the matter of level of development, the NJPC project affected area was placed in a high level of development in terms of sex ratios, literacy percentage, female literacy percentage, literacy percentage among scheduled caste and scheduled tribes populations, female literacy among SC/ST population, percentage of main workers to total population, percentage of female main workers to total female population and the percentage of families placed above the poverty line. Measurement of development index in all the districts of Himachal Pradesh including NJPC project affected area revealed that in terms of overall socio-economic development the NJPC project affected families were placed at the top position.

The principle most generally invoked with regard to resettlement and rehabilitations that compared with the pre-displacement situation, all project affected persons should have a similar or higher standard-of-living. This principle has a broader thrust that the cash-for-land or the land-for-land principles of compensation. Those may be even thought of as means to implement this general principle.

2.3 HYDROELECTRICITY AND RURAL DEVELOPMENT

Scholarly studies on rural electrification in India, as even on rural energy, are not only limited but also of relatively restricted scope. That is rather unusual as the transformative potential of electricity has been widely established and there is a near consensus of it across ideology, class and space. Most available literature either deal with the pattern of consumption, availability and distribution of electricity in rural areas or treat electricity as a component of the rural infrastructure while examining the likely impact of infrastructural investment upon regional economic development, or income generation to be specific. So far as primary data-based studies on rural energy are concerned, the major findings include: (a) the household sector, mainly cooking, have been the largest consumer of energy (as compared to the farm sector), although electricity forms a tiny proportion of all the energy consumed; (b) there has been a rise in the use of energy by both the farm and transport sectors; and (c) inability to access affordable energy has contributed importantly to poor quality of life in rural regions.

Even while exclusive studies on the impact of electricity on rural development in the Indian context are hard to come by, a few exercises have held out electricity as one of the key infrastructural components which contributes to agrarian and rural progress. Drawing upon these studies, and wider literature on rural infrastructure, Figure 2.2 provides a comprehensive picture of contemporary linkages between electricity and rural economy. It needs to be noted that while the four broad sectors/sites may have greater linkage possibilities, only those activities/functions have been noted which are widely observed given the current techno-economic limits in the rural context. In fact, so far as alternative energy sources available to the Indian villages are concerned, electricity-driven appliances/machines are still quite restricted; a great variety of agricultural and industrial activities including heating, drawing water and preliminary processing of crops are still done using human or animal power and other forms locally available energy as bio-based ones. That essentially implies that the use of electricity for a variety of existing and new purposes would crucially hinge upon the overall growth of basic social and economic infrastructure, natural endowments as also its nature and extent of interconnectedness with the modern urban-industrial dynamic (Das, 2006:2).

FIG. 2.2
Electricity and Rural Development Nexus

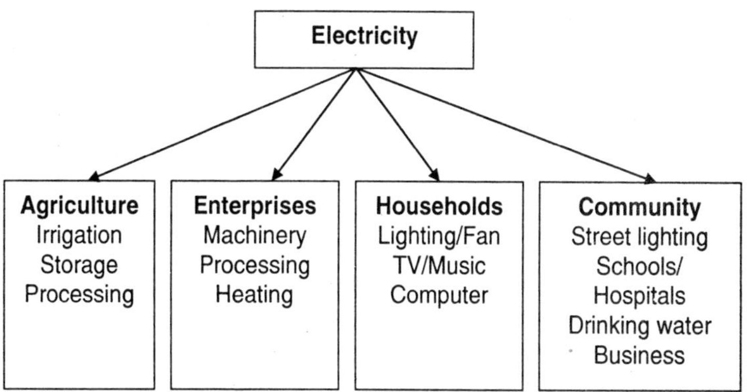

Both national and global experiences have amply suggested that there exists close correlation between the lack of access to electricity to electricity and rural poverty. Evidently, electricity has been viewed as a pre-requisite for both improving the standard of living and for carrying out productive and economic activities. Electricity as an energy input for essential functions such as drawing water for domestic consumption and irrigation, lighting that extends working and learning hours; powering tiny and small scale rural enterprises; and facilitating community level activities confers substantial benefits on the rural society. In fact, the positive impacts of electricity are "considerably greater due to a bundling of socio-economic benefits (TERI, 2002:2). It entails not only the development of rural areas but also the development of rural people as well. Our country has continental dimension and the pattern of rural development varies from area to area. It was during the Second Five Year Plan that the awareness towards the imbalances in rural were initiated. The rural development involves structural changes in the socio-economic situation to achieve improved living standard of low income population residing in rural areas and making the process of their development as more self-sustained. It includes economic development with close integration among various sections and sectors; and economic growth specifically directed to the rural poor.[3] In fact, it requires area-based development as well as beneficiary-oriented programmers. Furthermore, the development of rural area depends on location of various economic and social activities, their integration and proper linkages within and outside the areas.

The socio-economic infrastructure plays a vital role in influencing the level and nature of economic and socio-cultural activities in a Nation (Tiwari, 2000:1). According to the World Development Report, 1994 the growth of infrastructure not only facilitates the achievement of the goal of economic growth but also helps in eradicating poverty and maintaining environmental sustainability. Infrastructure can deliver major benefits in economic growth, poverty alleviation and environmental sustainability (World Development Report, 1994). The economic infrastructure usually take the form of physical capital formation and may include the long lasting engineering structure, equipment and facilities and the services they provide that are used in economic production and by households. It takes shape in the form of public utilities like power, piped gas, telecommunications, water supply, and public works such as major dam and canal works for irrigation. Hence physical and economic infrastructure and its services play a key role in quickening and impeding the pace of development (Tiwari, 2007). The heart of infrastructure is link perhaps the generation of electricity and its supply bridges this gap.

Ministry of rural development is organized in bringing rapid sustainable development and socio-economic transformation in rural India. It is important for economic liberalization, structured adjustment, safety network and allocation of resources. Various programme has been launched, e.g. Sampoorna Grameen Rozgar Yojna (SGRY), New Program for National Food for Work Program (NFFWP), Pradhan Mantri Sadak Yojna (PMSY), Swaraj Gram Swarojgar Yojna (SGSY), Drought Prone Area Programme (DPAP), Desert Development Program (DDP), Integrated Wastelands Development Program (IWDP); Accelerated Rural Water Supply (ARWSP), Central Rural Sanitation Program (CRSP), Total Sanitation Campaign (TSC), and NFFWP (National Food for Work Program), for maintaining sustainable development. There is Electricity Act, 2003, Appellate Tribunal for electricity, National electricity policy, Rural Electrification Corporation Ltd. (REC) and Power Finance Corporation that are playing very important role in the electrifying rural areas. The rural electrification programs involve providing electricity to the villages and other production-oriented activities such as minor irrigation, and rural industries. During 1950-51 there were only three thousand villages electrified in India that abruptly increases and found 507 thousand in 2001-02. There were 0.02 million pump sets energized during 1950-51 that were found 6 million in 1984-85 and 12 million in 2001-02 and thereafter continuously increasing day-by-day (Planning Commission, 10th Five Year Plan). Total villages by March 2005 were 5,87,258 and electricity has reached at 4,98,877 villages. There are 1,43,76,081 pump sets that were energized up to March 2005. Beside this there were total 162 projects with 47,930 installed capacity prepared in 2004.

For increasing productivity of agriculture in rural areas the natural resources of hydroelectric power is being renewed. During 1950-51 its generation was only 560 MW that was increased in 2003-04 up to 29,500 MW. Hydel projects take long period for gestation as compared to thermal. It involves supply of energy for two types of program namely for the production oriented activities, e.g. minor irrigation, rural industries and for rural electrification. The rural electrification programs are formulated and executed by SEBS or the State Power Department under rural electrification program. The 49,887 villages of 5,87,258 villages in India were electrified up to March 2005.

To give impetus to the rural electrification, the government would pay special attention for creation and augmentation of Rural Electricity Distribution Backbone and Village Electricity Infrastructure so as to cover all un-electrified villages and rural households within a span of five years. Rural Electric Supply Technology mission (REST) has been set-up to oversee the implementation of schemes under AREP (Accelerated Rural Electrification Program). Programs of rural electrifications are provided financial assistance by REC or the Rural Electrification Corporation. Power Finance Corporation (PFC) provides term finance to projects in power sector. Central Power Research Institute (CPRI) and National Power Training Institute (NPTI) are also under the administrative control of Ministry of Power. A power-trading corporation has also been incorporated primarily to support mega power projects in private sectors by acting as single entity to enter power purchase agreement.

There is a strongest focus on hydropower to meet projected power requirement by 2012. Installed power generation in 1947 was 1400 MW that has increased up to 1,18,419.09 on 31.03.2005 (80,902.45 thermal + 30953.63 MW hydro, 3811.01 MW wind and 2770 MW Nuclear). The Central Electricity Authority (CEA) is a statutory organization constituted under section B(1) of Electricity Supply Act, 1948 which has been superseded by section 70(1) of Electricity Act, 2003, playing important role in formulating policies and programmes for power development in the controlling, planning and coordinating various developmental activities in the power sectors.

CEA advises Central/Government on the matter relating into National Electricity Policy formulated short-term and perspective plans for the development of electricity system and co-coordinated activities of planning agencies for optimal utilization of resources to sub-serve the interest of National economy and to provide reliable and affordable electricity for all consumers. CEA under Electricity Act makes regulation and standard (e.g. construction of electricity plants, electric line, connectivity to grid, concurrence of hydroelectricity scheme) for promoting human quality of life.

It is responsible for concurrence of hydropower development schemes of the Central/State/Private sectors for best development of river on its tributary for power generation, consistent with requirement of drinking water, irrigation, navigation, and flood control. It promotes and assists timely completion of schemes and projects for improving and augmenting electricity. It also identifies bottleneck and problem areas, initiations and remedial actions. It collects data concerning general transmission, trading, distribution and utilization of electricity and carry out studies relating to cost efficiency, competitiveness. Further, it advises central government, state government and regulatory commission as all technical matters relating to generation, transmission and distribution of electricity. It also advises state government licensees for generating, operating and maintaining electricity system.

The CEA promote integrated operations of regional grid system and facilitate exchange of power within the country from surplus to deficit region with neighboring countries for mutual benefits. It also promotes research in the matter affecting the generation, transmission, and distribution and trading of electricity. It makes significant contribution to the number of professionals in India as well as abroad, e.g. Conferences International Development Grants Research Electrification CIGRE, Board of Irrigation Power (BIS), Central Board of Irrigation and Power (CBI &P), CEA render consultancy services in planning and designing hydro, thermal and transmission projects.

The Appellate Tribunal for electricity established by the Central Government on 7th April 2004 whose National Electricity Policy notified by Government of India on 12 February 2005 is important for providing policy guidance to the Electricity Regulatory Commission and to the CEA for preparation of National Electricity Plans. Government has launched Accelerating Power Development Program (APDP) that upgrades sub-transmission and distribution system and improvement in commercial viability of the State Electricity Board by reducing the aggregate technical and commercial losses to around 15% against 50%. APDP has two component named as investment and incentive components that play very important role in sustainable development. In 1991 CEA directed that no private power project be considered for inviting leader and International Competitive bidding. Some projects where MoU and registration is feasible should be awarded for execution. The National Thermal Power Corporation was incorporated in November 1975 within the objective of planning, promoting, organizing and integrating development of thermal power in the country. NTPC is a wholly owned subsidiary to carry out business of implementing and operating small and medium hydropower projects. It has signed MoU with REC for setting up the decentralized distribution generation scheme for rural electrification through non-conventional energy sources. Vidyut Vyapar Nagar (VVN) and subsidiary of NTPC transacted its business with State utility trading in more than 2613 MU in 2004-05.

In India such development is as a result of the vision of our great leader like Pt. Jawahar Lal Nehru who always insisted that science and technology should play important role in the area and the Dams should be considered as the temple of India. But, once a time, the innovative technology in the field of generation of electricity was not so satisfactory and praiseworthy in India that in contemporary scenarios are speeding up by such concept. It can be witnessed from the statement of our Former Prime Minister Lt. Sh. Rajiv Gandhi who was addressing to the State Irrigation Minister in 1986 expressed his concerns that *"The situation today is that since 1951, 256 big surface irrigation project have been initiated and only 65 completed and 181 under construction. This is not a happy state of affairs, we need some definite thrust from the projects that we started after 1970 perhaps we can safely say that almost no benefit has come to the people from these projects for 16 years"*. We have poured many out; people have got nothing back, nor irrigation, no water, no increase in production and no help in their daily life despite by pouring money out to a few contractors. In Himachal Pradesh approximately 91.3% people lives in villages. Purpose of rural electrification is to lighten as well enlightened the rural peoples of the villages. In Himachal Pradesh commendable efforts have been raised by the Government for launching various hydel projects that itself can be reflected by below mentioned table, proving as worth for raising economy and standards of living.

2.4 PLANNING FOR POWER DEVELOPMENT IN INDIA

Considering the importance of electricity in the process of economic development, due emphasis has been laid on power development in India right from the beginning of the planning era. During the First Five Year Plan, power and transport sectors received the largest share within economic infrastructure. But the expenditure on transport and communication was about half (46 per cent) of the total plan expenditure on economic infrastructure followed by irrigation and flood control (about 39 per cent). The power sector has been given second highest priority from the Third Five Year Plan onwards. The most notable shift which took place after the Third Five Year Plan was relative increase in the expenditure for power development. The expenditure on power development was about one-third of the total expenditure on economic infrastructure sector during the Third Plan, which marginally increased both in relative and absolute terms in the subsequent plans. During the Fourth Five Year Plan, expenditure on power sector increased by 134 per cent as compared to the Third Five Year Plan and further by 142

TABLE 2.3
Planning for Power Development in India

(*Rupees in crore*)

Sl. No.	*Plan*	*Total Plan Investment*	*Investment on Power*	*% of Power Investment to Total*
1.	First (1951-56)	2378	260	11.1
2.	Second (1956-61)	4800	427	8.9
3.	Third (1961-66)	8577	1253	14.6
4.	Fourth (1969-74)	15779	2932	18.6
5.	Fifth (1974-78)	39462	7100	18.8
6.	Sixth (1980-85)	109291	19264	19.8
7.	Seventh (1985-90)	218729	34274	19.1
8.	Eighth (1992-97)	434100	115561	26.6
9.	Ninth (1997-2002)	859200	222375	25.9
10.	Tenth (2002-07)	1525639	403927	26.5

FIG. 2.3
Investments on Power Sector under Plans

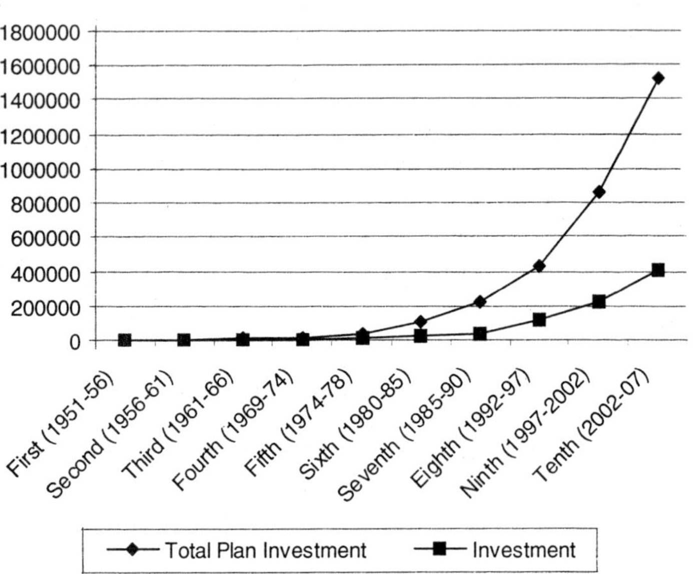

per cent in the Fifth Plan, in the Sixth Plan by 171 per cent, and by 78 per cent and 237 per cent during the Seventh and Eighth Plans respectively. During Ninth Plan a capacity addition of 46,820 MW was made while targeted capacity was only 40,250 MW consisting of 11,910 MW in Central sector and 10,750 in State sectors and 17,590 MW in private sectors.

2.4.1 Demand Supply Gap in Power Sector in India

Amongst the important sub-sectors, Power sector has shown significant gaps. Table lists the demand supply gap in the power sector in the 1990s. The poor performances of SEBs, with increasing financial strain emanating from low average tariffs and high cross subsidies to agriculture and household sectors has stifled the growth of this sector.

In the Telecommunications sectors, while the gap has significantly narrowed from about 27.9% in 1991-92 to 12.2% in 2000-01, reflecting the impact of reorientation of the policies

TABLE 2.4
Demand Supply Gap in Power Sector in India

(in billion KWh)

Year	*Requirement*	*Availability*	*Deficit*	*Deficit as % of Requirement*
1990-91	268	247	21	7.9
1991-92	289	266	23	7.8
1992-93	305	280	25	8.3
1993-94	323	299	24	7.3
1994-95	352	327	25	7.1
1995-96	390	354	36	9.2
1996-97	413	366	48	11.5
1997-98	425	390	34	8.1
1998-99	447	420	26	5.9
1999-2000	480	451	30	6.2
2000-01	507	467	40	7.8

Source : Ministry of Power, Government of India.

FIG. 2.4
Demand and Supply Gap in Power Sector in India

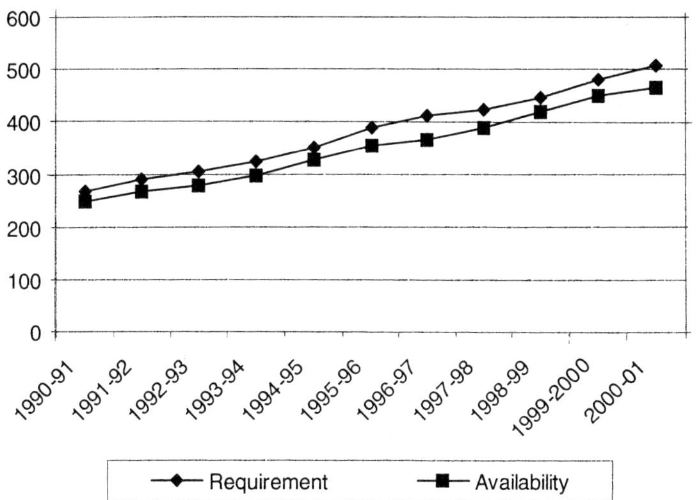

followed in the sector, it is still higher than the power sector. With the public finances totally inadequate to meet the potential demand, a large-scale entry of the private sector is essential to bridge these sectoral gaps.

While several initiatives have been made to attract private investment in infrastructure development, and there has been some success, the potential so far developed, is far short of that required. A major problem coming in the way of private sector participation is the rationalization of user charges with many infrastructure sectors suffering from the levy of inadequate user charges. Bankable investment can not be made in infrastructure projects unless they can be financed by the levy and collection of appropriate user charges.

2.4.2 Planning for Power Development in Himachal Pradesh

Considering the importance of power development, especially the hydel power development because of advantageous geographical features of the State, the government of Himachal Pradesh has been accorded top priority from Sixth Plan onwards. The share of plan investment in the power sector is depicted in Table 2.3. It may be observed from Table 2.3 that power sector has remained in the priority sector under Five Year Plans. Consequently, the installed capacity which was pitiable at 2.000 MW in 1950-51 has increased to 48.919 MW in 1970-71, and further to 126.520 MW in 1980-81 and 326.300 MW in 2002-03. The growth rate of the installed capacity was very responsive. The installed capacity of power sector has expanded by 163 times from the installed capacity that existed during 1951. Accordingly, the percentage of villages electrified to the total villages increased from 0.5 per cent in 1950-51 to 97.86 per cent in 2005-06. The number of agricultural pump sets energized also increased significantly by 173 times during this period. Per capita consumption of power was only 0.90 KWh (whereas all-India average was 17.8 kWh) in 1950-51 which remarkably rose to 339.0 kWh in 1999-2000 which was slightly higher than all-India average of 335 kWh.

TABLE 2.5

Planning for Power Development in Himachal Pradesh

Sl. No	*Plan*	*Total Investment*	*Investment on Power*	*% of Power Investment to Total*
1.	First (1951-56)	527.25	21.59	4.09
2.	Second (1956-61)	1602.60	150.69	9.40
3.	Third (1961-66)	3384.47	240.14	7.10
4.	Annual (1966-67)	946.05	295.04	31.19
5.	Annual (1967-68)	1443.94	395.52	27.39
6.	Annual (1968-69)	1595.19	415.75	26.06
7.	Fourth (1969-74)	11342.97	2450.03	21.60
8.	Fifth (1974-78)	16148.48	4053.89	25.10
9.	Annual (1978-79)	6810.17	1248.54	18.33
10.	Annual (1979-80)	7945.36	1550.00	19.31
11.	Sixth (1980-85)	65566.00	17924.95	27.34
12.	Seventh (1985-90)	132475.75	34747.61	26.23
13.	Annual (1990-91)	37762.93	6721.58	17.80
14.	Annual (1991-92)	40740.00	5344.34	13.12
15.	Eighth (1992-97)	349905.00	67692.79	19.35
16.	Ninth (1997-2002)	570000.00	101965.00	17.88

Source : Compile from Five-year Plan (various issues), Planning Department and Government of Himachal Pradesh, Shimla.

2.4.3 Generation and Consumption of Electricity in Himachal Pradesh

Table 2.6 shows the progress achieved in the production and consumption of hydropower in absolute terms. The analysis of generation of electricity has special significance in the present context because Himachal Pradesh happens to be one of the richest areas in the country in the matter of hydro-electric potential, as already mentioned in this chapter. Before the period of attainment of full-fledged statehood, i.e. 1971 the generation of electricity was negligible as is clear from Table 2.6. There was initially a relatively slow growth of both production and consumption of electricity until the late sixties. However, since the early seventies there has been a spectacular rise in both

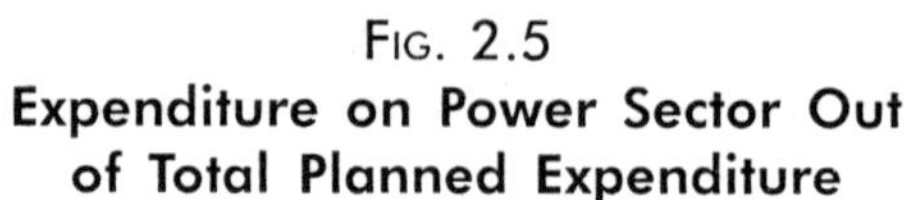

FIG. 2.5
Expenditure on Power Sector Out of Total Planned Expenditure

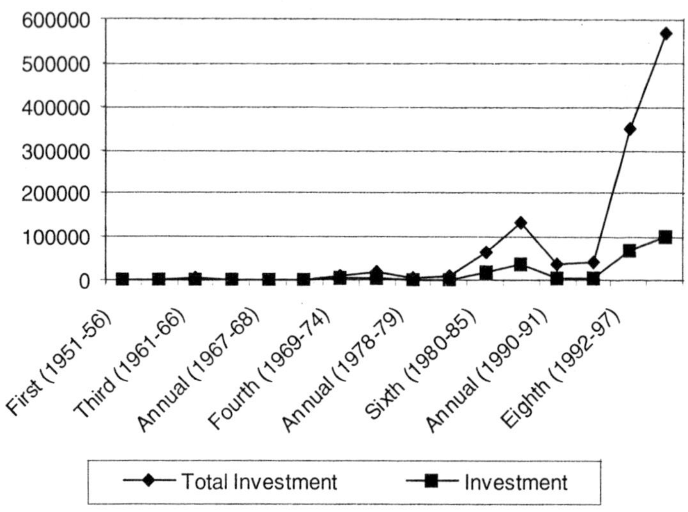

these respects. The performance in this regard has been quite marked since the end of the Third Plan, as can be seen from Table 2.6.

It may be observed that power generation which was 2450.66 MW in 1980-81 touched the level of 12623.79 MW in 1990-91. The short fall in over-all generation during 2001-02 is mainly due to less snowfall and scanty rainfall. The total electricity generation in the year 2003-04 touched the level of 13569.50 MW. During the thirty-three year period between 1970-71 to 2003-04, while electricity generated increased by as much as 26 times, electricity consumption also increased by more than 24 times. Further, the details of sale of power and consumption of power have been presented in Annexure I and II respectively. It is evident from the listings of these annexures that in absolute terms both sale and consumption of power have increased considerably.

The debate about strong relationship between hydroelectric generation and economic development has been settled. Sustainable economic development and achieving higher

Table 2.6
Generation and Consumption of Electricity in Himachal Pradesh

Year	*Electricity Generated (in lakh kWh)*	*Electricity Consumed (in lakh kWh)*
1950-51	3.58	9.97
1955-56	8.71	9.73
1960-61	10.71	25.75
1965-66	21.64	138.68
1970-71	528.41	1119.64
1975-76	1822.41	2205.43
1980-81	2450.66	2647.34
1985-86	5867.88	5633.16
1990-91	12623.79	10087.43
1995-96	11316.80	13396.82
2000-01	11533.00	22058.00
2001-02	11495.00	23318.00
2002-03	12773.00	22165.00
2003-04	13569.50	27263.24

Source : Compiled from Five Year Plans (Various issues), Planning Department, Government of Himachal Pradesh.

standards of living or quality of life depend upon the availability of adequate and reliable power at an affordable price. Unlike other commodities, electricity cannot be stored for future use. In other words, its generation and consumption have to be simultaneous and instantaneous. Furthermore, electricity, a major source of energy, is regarded as an important factor for bringing about radical change in the socio-economic condition of the masses. Because of multifarious uses of electricity, such as for lightning and as a source of motive power, its introduction does not merely facilitate provision of better amenities but augments productive capacity in different sectors of the economy through its wide range of applications. The availability of electric power affects industries in two principal ways. Firstly, the traditionally operated industries gradually

FIG. 2.6
Generation and Consumption of Electricity in Himachal Pradesh

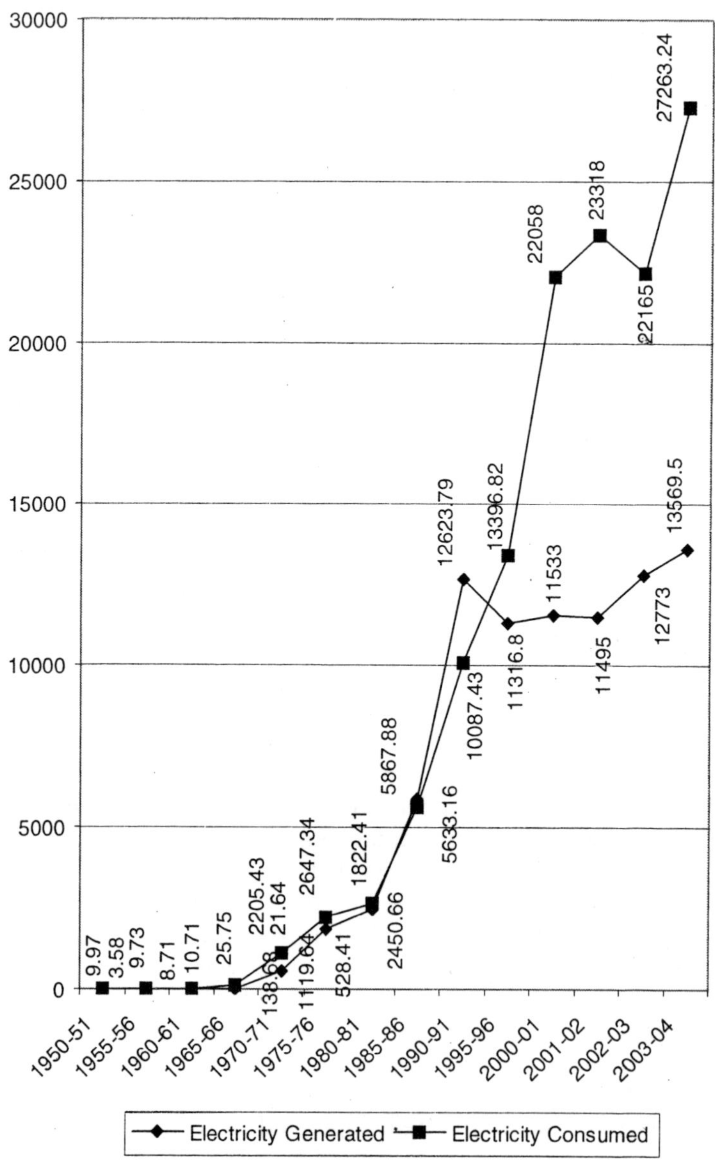

change over to the use of electric power, and secondly, new industrial units operated by power come up. Use of electric power, therefore, strengthens and expands both small and large industrial sectors and makes them more efficient and productive. Similarly, in the rural areas where economy is agricultural based, the role of electricity has been realized in the shape of increased productivity of land and labour. Inadequate power supply has remained one of the major factors of low productivity in the agricultural sector. The services sector has also been progressing with the progress in the generation of power.

The successive chapter 3 is intended to study of role of hydroelectric projects on the River Satluj Valley, particularly projects under execution under Satluj Jal Vidyut Nigam (SJVN) in improving quality of life as well as socio-economic development of the state of Himachal Pradesh in general and in the areas of execution of these hydroelectric projects in particular.

2.5 MAIN HIGHLIGHTS

- Energy is an essential input for economic development and it also facilitates in improving the quality of life of the general population. Development of conventional forms of energy for meeting the growing energy needs of society at a reasonable cost is responsibility of the Government, Development and promotion of non-conventional/alternate/new and renewable sources of energy such as solar, wind and bio-energy, etc., are also getting sustained attention. Nuclear energy development is being geared up to contribute significantly to the overall energy availability in the country.
- Development of hydroelectric power direct towards increased provision and use of energy services, which is an integral part of enhanced economic development. Advanced industrialized societies use more electric power per unit of economic output and far more electricity per capita than poorer societies, especially those still in a pre-industrial state.

Electricity use per unit of output does seem to decline over time in the more advanced stages of industrialization, reflecting the adoption of progressively more proficient technologies for energy production and utilization as well as change in the composition of economic activity. Electricity intensity in developing countries perhaps peaks faster and at a lower level along the development path than was the case during the industrialization of the developed world. But even with trends toward greater energy efficiency and other dampening factors, total energy use and energy use per capita continue to grow in the advanced industrialized countries, and even more rapid growth can be expected in the developing countries as their incomes advance. Development involves a number of other steps besides those associated with energy, notably including the evolution of education and labour markets, financial institutions to support capital investment, modernization of agriculture, and provision of infrastructure for water, sanitation, and communications. This is not just an academic question; energy development competes with other development opportunities in the allocation of scarce opportunities for policy and institutional reform.

- Sustainable economic development and achieving higher standards of living depend upon the availability of adequate and reliable power at an affordable price. Unlike other commodities, electricity cannot be stored for future use. In other words, its generation and consumption have to be simultaneous and instantaneous. Furthermore, electricity, a major source of energy, is regarded as an important factor for bringing about radical change in the socio-economic life of a community. Because of multifarious uses of electricity, such as for lightning and as a source of motive power, its introduction does not merely facilitate provision of better amenities but augments productive capacity in different sectors of the economy through its wide range of applications. The availability of electric power affects industries in two principal ways. Firstly, the traditionally operated industries gradually change over to the use of electric power, and secondly, new industrial units operated by power come up. Use of electric power, therefore, strengthens and expands both small and large industrial sectors and makes them more efficient and productive

- The assessment of extent of relationship between hydroelectricity and economic development has shown positive and significant association between pairs of indicators, viz. Net State Domestic Product and Hydropower Generation; Net State Domestic Product and Consumption of Power; Per Capita Income and Hydropower Generation; Per Capita Income and Consumption of Power; Hydro-Power Generation and Industrialization; and Hydro-Power Consumption and Industrialization, etc.

Notes and Refrences

1. For detail, see Tiwari, A.K (2007): *Hydro Electric Power Generation and Economic Development (A Case Study of Nathpa Jhakri Hydro Electric Project of Himachal Pradesh)*, Project Report, submitted to the Institute of Himalayan Studies (UGC Centre of Excellence), Himachal Pradesh University, Shimla.
2. Power generated in Himachal Pradesh can become a significant catalyst and driving force for economic development not only for this state but for the entire northern region of the country. Power development has related aspect in the context of hill area development such as creation of work opportunities not only in the hydroelectric projects but in diverse-related activities like road construction, housing, transportation, etc. Even the trade and business activities that come up at the project site provide alternative job opportunities. These job opportunities are open mostly to unskilled or semi-skilled labour force.
3. A distinct outcome of the Indian development trajectory, with Plans and Reforms has been the downright sidelining of the rural infrastructure, excluding the repeated claims. Even the much-hailed and first-ever Indian Infrastructure Report, brought out in 1996, hardly had touched upon the whole gamut of issues concerning rural infrastructure. Scholarly studies on rural electrification in India, as even on rural energy, are not only limited but also of relatively restricted scope. That is rather unusual as the transformative potential of electricity has been widely

established and there is a near consensus over it across ideology, class and space.

4. The adequate supply of electricity has played a significant role in the socio-economic development of Himachal Pradesh. In the beginning of the planning era, the economy of Himachal was fundamentally backward. Now, the socio-economic status and quality of life in the state of Himachal Pradesh is now comparable with the developed states of the country. The electricity consumption in Himachal Pradesh, in comparison to some of the similar placed/neighbouring states and also with country's average, has enhanced significantly as shown in Table below:

States	*Households with Electricity (%)*	*Per capita consumption of electricity (kWh)*	
	2006	*1980-81*	*2000-01*
Himachal Pradesh	97.86	66.00	339.00
Jammu & Kashmir	71.55	75.00	268.00
Punjab	96.44	304.00	921.00
India	64.09	121.00	355.00

Source : Market Skyline 2006, Planning Commission, Statistical Abstract of India.

3

Hydroelectric Power Projects on the Satluj River Valley

In the proceeding chapters, hydroelectric scenarios, hydroelectric power generation and economic development have been examined. This chapter has five significant objectives. First, it provides an overview of the Satluj Jal Vidyut Nigam Limited. Secondly, it focuses on the resettlement and rehabilitation policy of SJVNL. Thirdly, it presents domino-effect of resettlement and rehabilitation policies. Fourthly, it provides general features of the villages affected by SJVNL Hydropower projects. Lastly, it investigates the role of hydropower projects on the Satluj River Valley on economic development with the help of Development Index.

The River Satluj, which is one of the key basins featuring in the hydro development plan of the state of Himachal Pradesh, rises in the Tibetan Plateau (Rakastal-Mansarovar lake; at an elevation of about 4570 m above mean sea level), travels about 1450 km (320 km in China, 758 km in India, and 370 km in Pakistan) before it meets the Chenab River and subsequently the Indus. Governments of Himachal Pradesh and India are working to exploit the full hydro-potential of the Satluj river

Basin though both private and public developers. Some of the projects proposed for construction on this river are Khab and the 1000 MW Karchham Wangtoo project upstream of Rampur and 425 MW Luhri and 800 MW Kol dam projects down stream. The 1500 MW Nathpa Jhakri HEP, immediate upstream is already in stage of operation. The most celebrated dam on the river is the Bhakra Dam, which was completed in 1963. Downstream of Bhakra too there are structures on the river, including the Nangal diversion dam and Ropar barrage.

The catchments area of the Satluj at Rampur is about 50,800 km^2 (49,800 km^2 at Nathpa Dam), of which about 30% falls in India and the remainder in China. The rivers in the catchments are fed by snow melt, particularly in China. A small portion of the project catchments also receive precipitation due to the South-West monsoon (June-September). The peak flows of the river occur during June to September, while the lean period occurs between October and April. Much hydrological study of the Satluj has previously been performed in preparation for the construction of the existing Bhakra Dam, which is downstream of Rampur, and for the construction of the upstream existing Nathpa Jhakri scheme. Water availability studies were carried out from 1963 onward by using observed discharges at Rampur town and river diversion works will be designed to withstand a river flow corresponding to a 10,000 years return period flood, which has been assessed to be 7,151 cubic meters per second at Rampur.

3.1 THE SATLUJ JAL VIDYUT NIGAM LIMITED: AN OVERVIEW

The Satluj Jal Vidyut Nigam Limited—SJVN (formerly Nathpa Jhakri Power Corporation Limited—NJPC) was incorporated on Mary 24, 1988 as a joint venture of the Government of India (GOI) and the Government of Himachal Pradesh (GOHP) to plan, investigate, organize, execute, operate and maintain Hydro-electric power projects in the river Satluj basin in the state of Himachal Pradesh and at any other place. The present authorized share capital of SJVNL is Rs. 4500 crore. The Nathpa Jhakri Hydro Electric Project (1500 MW) was the first project undertaken by SJVN for execution.

The 1500 MW, Nathpa Jhakri Hydro Electric Project—NJHEP (the largest underground hydro electric power Project in the country) was the first project undertaken by Satluj Jal Vidyut Nigam Limited (SJVN) for execution and has since been commissioned, during the 10th Plan, progressively between September 2003 to May 18, 2004. Besides the social and economic upliftment of the people in its vicinity, the 1500 MW NJHPS has been designed to generate 6950.88 MU of electrical energy in a 90% dependable year with 95% machine availability. It is also providing 1500 MW of valuable peaking power to the Northern Grid. Out of the total energy generated at bus bar, 12% is supplied free of cost to the home state, i.e. Himachal Pradesh[1]. From the remaining 88% energy generation, 25% is supplied to H.P. at bus bar rates. Balance power has been allocated to different states/UTs of Northern Region by the Ministry of Power, Government of India.

Apart from the above, indirect benefits have also accrued to the region by way of increase in agriculture and industrial production. As per the study conducted by Agro-Economic Research Centre, Himachal Pradesh University, the people and the areas in the vicinity of the NJHPS, have witnessed tremendous growth and upliftment, which has resulted into an increased literacy rate, average annual household income, level of employment and standard of living, etc. In addition, the project has provided gainful employment to a large number of skilled and unskilled workers and has also opened the landlocked hinterland by providing essential facilities such as schools, hospitals, etc. for the people of the area. Thus, NJHPS has ushered in social and economic upliftment of the persons living in the vicinity of the Project, i.e. of society at large.

SJVN entered the august club of selected Central Public Sector Undertakings (CPSUs) with the signing of first Memorandum of Understanding (MoU) for the year 2004-05, with the Ministry of Power, Government of India. Thereafter the MoUs for the year 2005-06, 2006-07 and 2007-08 have also been signed, wherein various targets have been set out for achievement during the years.

The Financial Performance of SJVNL, for the Last 3 Years

(*Rs. in Crores*)

Sl. No.	*Description*	*2006-07*	*2005-06*	*2004-05*
Income Details				
(i)	Sales	1618.23	1371.50	1098
(ii)	Other income	47.53	20.28	22.72
(iii)	Total Income (i + ii)	1665.76	1391.78	1120.99
(iv)	Profit before interest, Depreciation and Finance charges	1525.41	1250.49	1049.09
(v)	Profit before tax	825.81	543.61	323.84
(vi)	Profit after tax	732.71	498.22	298.43
(vii)	Dividend (excluding Tax on dividend)	235.00	159.43	143.15

Source : Annual Reports, SJVNL.

3.1.1 Environment

Satluj Jal Vidyut Nigam has been fully aware of the importance of both environmental, resettlement and rehabilitation issues. It has adopted an environment; resettlement and rehabilitation policy which reiterates company's commitment to sustainable development is within the carrying capacity of the eco-system and promotes the improvement of the quality of life. Initiative are taken to undertake steps for mitigation of the grievances of the project affected persons with the involvement of State Governments and to put into effect the approved remedial action plan and also for infrastructural development in the project areas.

Satluj Jal Vidyut Nigam believes that employees are its most valuable assets and has evolved growth-oriented Human Resource Development strategy. Empowerment of manpower skills through training receives utmost importance every time. The company has well established strategy for imparting training to the employees. Training imparted is two dimensional i.e., in-house training and through external professional institutions as well. The company also facilitates the professional candidates of various institutions for undergoing vocational training in the organization.

Satluj Jal Vidyut Nigam Limited has established Hydel Training Institute at Kotla which is about 14 Kms from Jhakri on Jeori-Sarahan road amidst beautiful natural surroundings. This institute was opened for imparting training in October 2003 and is functioning effectively and providing training to SJVN employees besides employees of other power sector organizations in the hydro power sector.

In order to ensure that implementation of the Official Language Policy of the Government of India, all possible efforts have been made by the Corporation to achieve the targets as specified by the Department of Official Language. The organization has bagged a number of Awards in recognition of its efforts to implement this Policy.

SJVN has the experience of corporate and project planning, design, engineering, construction management, erection and commissioning, contracts management, project management, human resource management, financial management and commercial management of India's largest hydro-electric project. To effectively utilize the in-house expertise and the experience gained, a dedicated consultancy division has been established for providing consultancy services to national and international organizations. The consultancy in the field of hydro power, road/railway tunnels, etc. has been provided to various organizations.

In order to enhance the quality of technical inputs/skills gained out of the experience of execution of NJHEP, a R&D Cell has been created to identify the areas in project execution and Operation & Maintenance, on which further R & D can be carried out so as to utilize the enhanced knowledge for execution/operation of the Power Plants in the country.

SJVN abides by its moral responsibility for maintaining a safer environment for all its employees and all those of its contracting agencies. Due attention is given to health and safety aspects in working areas. The safety measures adopted encompass the best codes and practices which are disseminated to all the employees for ensuring compliance at all level.

SJVN achieved the distinction of having been awarded the prestigious ISO 9001:2000 certification from M/s. Det Norske Veritas (DNV), one of the world's most reputed certification body, in recognition of its note-worthy efforts for well maintained quality management systems. SJVN was declared an ISO 9001:2000 company w.e.f. February 02, 2005. It has also been declared as ISO 14001:2004 Company on July 28, 2006, for environmental management systems of its Nathpa Jhakri Hydro Electric Power Station, which is one of the most environment friendly hydro power projects in the world. It has also obtained ISO 9001 for operation and maintenance of its Nathpa Jhakri Hydro Electric Power Station on February 02, 2007.

The Rampur HE Project is being located on the river Satluj in the Shimla and Kullu districts of Himachal Pradesh is envisaged as a run-of-the-river Project with minimal social and environmental impacts. This project will require neither a Dam nor any new reservoir capacity, nor will it lead to any land inundation, because it is being essentially envisaged as cascade station to SJVN's existing 1500 MW Nathpa Jhakri Power Station (NJHPS).

The 412 MW Rampur Hydro Electric Project (RHEP) is a tailrace development of Nathpa Jhakri Hydro Power Station (NJHPS). A tiny reservoir in the shape of pond of the tailrace outfall of NJHPS connects the two projects. The intake of the RHEP, which stands constructed as a part of the TRT outfall shall draw water from this pond.

As the Rampur HE Project scheme will utilize water existing from 1500 MW NJHPS Tail Race Tunnel at village Jhakri (District Shimla), it will not require any additional De-silting Chambers, the water being already de-silted within the NJHPS Plant. From the intake, a 15 Km underground tunnel will conduct the de-silted water to a surfacfe Poer House at village Bael (opposite to village Duttnagar in district Kullu). The Project's six 68.67 MW turbines will generate 1770 million units of power (in a 90% hydrologically dependable year); and this electricity will be used in the states of Northern Region of India. Twelve percent of the power generated will be provided free of cost to the Govt. of Himachal Pradesh as royalty. Subsequent to clearance of Cabinet Committee on Economic Affairs, accorded by Govt. of India, SJVN has awarded major civil works on February 01, 2007, and the construction works are in full swing. Tentative date of commissioning of 412 MW Rampur HE Project is January, 2012.

SJVN, contemplates to become a 6000 MW company within ten years for which a number of projects in the states of Himachal Pradesh, Uttarakhand, and Arunachal Pradesh have been taken up for preparation of DPR and subsequent execution (*Times of India*, New Delhi, Chandigarh, February 2, 2009):

In The State of Himachal Pradesh

(1) Luhri Hydel Project (775 MW) on river Satluj, in district Shimla of Himachal Pradesh, India.
(2) Khab HE Project (1020 MW) on river Satluj, in district Kinnaur of Himachal Pradesh, India.

In The State of Uttarakhand

(1) Devsari Dam Hydel Project (252 MW), on river Pindar, located in district Chamoli, of Uttarakhand, India.
(2) Naitwar Mori HE Project (34.5 MW), on river Tons (a tributary of river Yamuna), located in district Uttarkashi of Uttarakhand, India.
(3) Jakhol Sankri HE Project (33 MW), on river Supin, located in district Uttarkashi, of Uttrakhand, India.

3.2 NATHPA JHAKRI HYDRO-ELECTRIC POWER PROJECT

Nathpa Jhakri Hydro-electric Project is being executed on river Satluj between villages Nathpa and Jhakri situated in Kinnaur and Shimla districts of Himachal Pradesh. It envisages harnessing of 1500 MW hydel power potential utilizing 405 comics of river water in the form of a run off river scheme by constructing 60.6 mtr. high concrete gravity dam, through four intakes and four underground desilting chambers. Nathpa Jhakri Power Corporation has acquired 123 hectare of forest land and 224 hectare of private land for this gigantic project.

The successful completion and commissioning of prestigious 1500 MW Nathpa Jhakri Hydro Power Project in Himachal Pradesh in September 2003 was an important landmark for the corporation. The project has already generated

12,386 million units (MU) (up to 20 June 2006) of electrical energy as per detail listed below:

Year	*Generation*
2003-04	1116 MU
2004-05	5108 MU
2005-06	4055 MU
2006-07 (up to June 20, 2006)	2107 MU

Source : Annual Reports, NJPC (Now SJVNL).

This project, being a run of river scheme, has had a minimum adverse impact on the environment thus maintaining the ecological balance in the project vicinity. This project was accorded environmental clearance by Department of Science and Technology in 1980. While according approval for the use of forest land, Ministry of Environment and Forest, Government of India imposed some stipulations to minimize the impact of the project on the environment. To fulfil these requirements NJPC has already funded State Forest Department for compensatory afforestation, soil stabilization and avenue plantation in the project vicinity. The 1500 MW Nathpa Jhakri Hydro-electric Power Project (NJHPP) is the first project commissioned by Satluj Jal Vidyut Nigam (SJVN). SJVN is also planning to take up more projects in the Satluj River basin in Himachal Pradesh. It has already taken up three more projects in Himachal Pradesh, Rampur HE Project (412 MW), Luhri HE Project (700 MW) and Khab HE Project (1020 MW). Beneficiaries of the project include States of Himachal Pradesh, Haryana, Rajasthan, Jammu & Kashmir, Uttar Pradesh, Uttaranchal and Delhi.

SJVN has paid a dividend of Rs. 35.40 crore to Government of Himachal Pradesh for its operations for the year 2004-05. From its nearly full year's commercial operations for the year 2004-05, the company has posted a net profit of Rs. 298.43 crore on a turnover of Rs. 1098.28 crore. The company has declared a maiden dividend of Rs. 143.16 crore for the year 2004-05 to be shared by the equity partners, viz. Government of India and the Government of Himachal Pradesh in the ratio of 3:1 respectively. The Company's flagship venture 1500 MW Nathpa

Jhakri Hydro Power Project had commenced commercial operations in October 2003 and has since been supplying valuable power to the northern grid beneficiary states viz. Delhi, Rajasthan, Uttar Pradesh, Punjab, Haryana, Himachal Pradesh, J&K and Chandigarh. During the year 2004-05 the Project had generated 5108 million units of electricity. SJVN is presently executing three more projects in Himachal Pradesh namely 1020 MW Khab Hydro-electric Project, 412 MW Rampur Hydro-electric Project and 700 MW Luhri Hydro-electric Project.

The inordinate delay in taking up the Rampur Project notwithstanding, the 412 MW plant will still be one of the most economical hydroelectric ventures. The project is essentially the stage-II of the 1500 MW Nathpa Jhakri project, the largest hydro-electric scheme of the country, which was commissioned in 2003 by the Sutluj Jal Vidyut Nigam (SJVN). Work on the project should have started much before the completion of the Nathpa Jhakri project but the state and the Centre, the two partners in the SJVN, failed to arrive at an agreement. In fact, an agreement was finalized in July 2002 but it could not be signed as employees of the State Electricity Board had opposed it (*The Tribune*, Chandigarh, Saturday, January 13, 2007, p. 9).

The Congress government, which came to power subsequently, had renegotiated the project to get its equity share raised from 25 to 30 per cent and include a provision that at least 70 percent manpower for the project will be recruited within the state. An agreement was finally signed in October 2004.

In the meantime, the cost of the project had escalated from Rs. 1734 crore to Rs. 2047 crore. The cost of generation, which was less than Rs. 2 per unit, has gone up to Rs. 2.39 per unit.

It is the most economical project as power will be generated by utilizing the de-silted tailrace water of the Nathpa Jhakri project. No big diversion dam or de-silting chambers will be required to be constructed. A 15-km-long head race tunnel (HRT) will carry the tail race water from the Jhakri to Bayal, 10 km from Rampur, where the power house will be constructed. It will start from the left bank and crossover to the right bank underneath the bed of the Sutluj from a point 484 in down stream Jhakri. The gross head is 138 m, while head available from power generation will be 119 m. Besides the tunnel, a 140-m deep surge shaft will also be constructed.

Unlike Nathpa Jhakri, which boasts of the largest underground power station of Asia, the power house will be constructed over-ground. As the power house will function in tandem with the Nathpa Jhakri project, as many as six turbines, each of 68.67 MW capacities, will be installed. The Nathpa Jhakri has six units of 250 MW each. In all 1770 million units of electricity will be generated.

The project will be financed on a 70 to 30 debt equity ratio and the SJVN will provide Rs. 614 crore from its internal resources. The state will also be entitled to 12 per cent of the power generated, free of cost, as royalty. The total requirement of land is 101 hectare out of which 69 hectare is forest land and the rest private land (*The Tribune*, Chandigarh, Saturday, January 13, 2007, p. 9).

3.3 RESETTLEMENT AND REHABILITATION POLICY OF NJPC

The NJPC has followed the resettlement and rehabilitation policy formulated in consultation with the Government of Himachal Pradesh to provide assistance and rehabilitation measures to all those who are affected by the project. Its provisions include:

- Developed agricultural land to landless PAFs (Project Affected Families) equivalent to the area acquired or 5 bighas whichever is less. This 5 bighas would include any land left with the family after acquisition. Allotment of land will be made on the basis of landless certificate issued by the SDM of Rampur.
- Substitute house with a plinth area of 45 square meter or a payment of Rs. 45,000 according to their choice, to each PAFs whose houses were acquired.
- Allotment of plots for shops at the Jhakri market for displaced shopkeepers.
- Preferences in allotment of shops at the shopping complex built at Jhakri to displaced shopkeepers and other PAFs.

- Provision of suitable employment to one member of each landless PAFs according to his capacity and qualifications, subject to availability of vacancy. However, a PAF who has been allotted a shop plot will not be eligible for employment.

NJPC had adopted the following environment resettlement and rehabilitation Policy in April 1997:

"We believe in sustainable development which is within the carrying capacity of supporting ecosystems and which caters to human needs and improves the quality of life. We are, therefore, committed to:

(a) Respect and care for the community of life.
(b) Bring about changes in personal attitudes and practices to enable the affected community to care for their own environment.
(c) Address legitimate concerns to project affected persons.
(d) Conserve the Earth's vitality and diversity.
(e) Minimize the depletion of non-renewable resources.

We will take utmost care to ensure that our activities do not threaten survival and quality of life of project affected persons by protecting, to the extent possible, their habitats, natural systems and resources, and minimize depletion of non-renewable resources. Our special emphasis will be to achieve our objectives through community participation and by treating PAPs fairly and in keeping with the laws of the land. We shall endeavor to encourage education, family welfare, role of women-in-development, energy conservation, and basic necessities".

Abstract of National Policy on Resettlement and Rehabilitation for Project family has been appended as Appendix-I for perusal. The policy has been framed in that manner so that the adequate compensation to the affected families could be provided. The SJVNL is a public sector organization, the national policy for Resettlement and Rehabilitation are fully implemented for the welfare of the project affected families. A detail description of R&R policy of various projects under SJVNL has been presented in the following paragraphs.

3.4. RESETTLEMENT AND REHABILITATION POLICY OF OTHER HYDRO POWER PROJECTS UNDER SATLUJ JAL VIDYUT NIGAM LIMITED

3.4.1 Resettlement and Rehabilitation Schemes for Rampur Project

- This scheme may be called as the Resettlement and Rehabilitation scheme of Satluj Jal Vidyut Nigam Limited for the project affected families of Rampur hydro electric project.
- It shall extend to the whole of area affected of likely to be affected as a result of construction of Rampur Hydroelectric project within Shimla and Kullu districts of Himachal Pradesh.
- The commissioner of Resettlement and Rehabilitation appointed by State Government for supervising the relief and Rehabilitation works of various projects in Himachal Pradesh would also be commissioner for Resettlement and Rehabilitation under this scheme. The resettlement and rehabilitation works shall be carried out under this direction and guidance.
- Deputy Commissioner, Shimla/Deputy Commissioner, Kullu in whose jurisdiction the project affected areas falls, will be the Administrators for Resettlement and Rehabilitation of the respective areas. The relief and rehabilitation work would be carried out and controlled under their supervision for areas falling within their respective jurisdictions.
- *Family:* It means husband/wife of the person who is entered as owner/co-owner of the land in the revenue record, their children including step or adopted children, grand children and includes his or her parents and those brothers and sisters who are living jointly with him or her as per entries of Panchayat

Parivar Register as on date of notification under Section 4 of the Land Acquisition Act, 1894.

Explanation: Only the Panchayat Parivar Register Entry as it stood on the date of Notification under Section-4 of the Land Acquisition Act, 1894 shall be taken into account for the purpose of separate family for re habilitation benefits as well as for consideration of employment.

- Project Affected Area/Affected zones means area as notified by the project authority where land is acquired for construction for any component of project, infrastructure, township, offices, construction facilities welfare facilities, etc. Unit for declaring Project Affected Area would be Panchayat/Revenue village. It means a family or person whose place of residence or other properties of source of livelihood are substantially affected by the process of acquisition of land for the project and who has been residing continuously for a period of not less that three years preceding the date of declaration of the Project Affected Area/Affected Zone or practicing any trade, occupation or vocation continuously for a period of not less than three years in the Project Affected Area/ Affected Zone, preceding the date of declaration of the affected zone.

Explanation: The date of declaration will be taken as the date of notification under Section 4 of Land Acquisition Act, 1894. The period of residence of not less than three years will not be applicable in respect of PAFs who own land in the Project Affected Area. The period of residence of not less than three years as well as effects on source of livelihood would be determined by the Deputy Commissioner concerned.

Project Affected Family Rendered Landless: The Project Affected Family rendered landless means that family whose whole agriculture land is acquired for the project or in whose case balance agriculture land left after acquisition is less than 5 Bighas. For this purpose agricultural land held within the project area by all such persons and their family members shall be reckoned. Person losing land on acquisition of building and land appurtenant there to shall not be treated as landless Project Affected Family. The landless PAF shall be certified by the Deputy Commissioner concerned.

- *Project Affected Family Rendered Houseless:* The Project Affected Family rendered houseless means the family whose dwelling house is acquired for the project. This will be certified by the Deputy Commissioner concerned.
- In addition to the above two categories there will be Project Affected families who will be rendered landless as well as houseless as per definition given above. They shall be eligible for benefits of project affected families rendered landless and project affected families rendered houseless. This will be certified by the Deputy Commissioner concerned.
- *Shopkeepers displaced by the Project*: Displaced shopkeeper means the shopkeeper who had taken shop on rent and had been genuinely running business therein as on date of issuance of notification under Section-4 of the Land Acquisition Act, 1894 and whose such shop is acquired for the project or the shop owner who was himself running his business in such shops.
- *Project Authority*: Project authority in normal connotation refers to project developer or project proponent e.g. State Government or Public Sector undertaking implementing a Project, etc. i.e. Satluj Jal Vidyut Nigam Limited in this case.
- Words and expressions used in this scheme but not defined here in shall have the same meaning as assigned to them in the Himachal Pradesh Nautor Land Rules, 1968.

3.4.2 Resettlement Grant for Rampur Project

The Project Affected Family which is rendered landless on account of acquisition of land shall be eligible for landless grant in the following manner:

- Families whose land before acquisition was more than 5 Bighas and is left with 1 Biswa or no agriculture land after acquisition Rs. 65,000.
- Families whose land before acquisition was les than 5 Bighas and left with 1 Biswa or no agriculture land after acquisition Rs. 55,000.
- Family whose land holding is left more than 1 Biswa and less than 5 bighas of land after acquisition Rs. 45,000.
- Family whose cattle shed is acquired in the project area shall get one time financial assistance of Rs. 5000. In no case the grant shall exceed Rs. 5000 per family.
- Each project affected family which is rendered landless as well as houseless (both) or houseless will be provided an independent house with a built up plinth area of 60 Sq mtr. Alternatively PAF can also be offered a plot of size which allows construction of built-up house of 60 sq. mtr. plinth are plus construction cost of the house @ 3000 per sq. mtr.
- A family which does not opt for house/plot but constructs his house at his on cost with a plinth area of 60 sq. mtrs. Or more upto 50% shall be paid the construction cost of the house @ Rs. 3500 per sq. mtrs. and options from such families will be asked at an appropriate time. In case any such family constructs house of less than 60 sq. mtrs. Plinth area on his own plot or plot offered by the Nigam then amount to be given will be worked out on prorate basis.
- Displaced shopkeepers will be given shops in allotment in the market complex of the Project Township wherever the Nigam constructs such market places. In addition they will be entitled to one time displacement grant of Rs. 10,000. The commercial premises/shops allotted to such displaced shopkeepers shall be utilized by them or their successors in the interests for bonafide use only. In case the Nigam is unable to provide shops, displaced shopkeepers shall get financial assistance of Rs. 40,000.

- Infrastructural facilities the rehabilitation colony will include water supply, sewerage, drainage, electricity streets and approach paths at the project cost.
- Transportation at the project cost will be provided for physical mobilization of all the PAFs and displaced shopkeepers as soon as the house shop get constructed in the Rehabilitation colony or a sum of Rs. 5000 in lump sum and the option will be invited from the affected families/shopkeepers.
- Stamp duty and other fees payable for registration shall be borne by the Project Authority. The Deputy Commissioner concerned will be the sanctioning authority for rehabilitation grant which shall be provided by the project authorities and placed at the disposal of the concerned Deputy Commissioner for disbursement to eligible concerned.
- All the above grants shall be in addition to the compensation paid under Land Acquisition Act, 1894.
- One member of each project affected family rendered landless will be provided for employment by the Project Authority in the category of skilled/semi-skilled/unskilled workmen subject to fulfilling the requisite criteria/qualification and as and when any fresh recruitment is done in these categories. It would be ensured that land oustees eligible for employment as mentioned above are given chance first and normal recruitment would be made only if none are available from amongst them. However, persons whose are allotted shops shall not be eligible for benefits of employment and *vice versa.*
- The following criteria will be adhered to by the Deputy Commissioner concerned for providing of preference while sponsoring the names for employment. Affected families whose entire land has been acquired. Affected families who have become landless on account of acquisition of land by the Nigam. Within these categories preference will be given on the basis of quantum of land acquired. Those who lose more land will come first.

- *Secondary employment:* There may be families who are not covered under the Project Affected family rendered landless/houseless/shop less but their land is acquired for the project shall have to be helped in starting some gainful occupation or getting training. Therefore, such families who may not be accommodated in direct employment, the Project Authorities will help them in any one of the following manners.
- Merit scholarship scheme for the wards of Project affected families who may be pursuing vocational or professional course shall be introduced by the project authorities as per scheme to be drafted by the Nigam in consultation with Govt. of Himachal Pradesh.
- The Project Authorities will also consider awarding petty contract to the co-operatives of eligible families on preferential basis so that some may be engaged in such jobs. Further, the project authorities will ad vise their contractors to engage eligible persons from affected families on the preferential basis wherever possible during construction stage.
- The project affected families will be assisted to start various suitable self employed occupations which include dairy, farming, poultry, weaving, bakery, handicraft, cottage, industries units/shops and hiring of vehicles to the corporation as per scheme to be drafted by the Nigam. The Project Authority will provide a grant of Rs. 30,000 per family towards seed capital. The grant will be given once only.
- Only those families who have not been provided with employment in the Nigam or have not have been allotted any shop will be eligible for this grant.

  Explanation: The deputy commissioner concerned will certify effect on source of livelihood in case of rural artisans, small traders and self employed persons for the eligibility of the grant.
- The Project Authority will provide support services fro Project Affected Families interested in horticulture, agriculture and veterinary.

- Free of cost cooker to each project affected family would be provided by the Satluj Jal Vidyut Nigam Limited.
- *Community Development:* Project affected areas/village after due assessment done by the committee constituted under the chairmanship of the Deputy commissioner will be provided with infrastructural up gradation schemes which will include Mobile health centre/Van, approach roads, internal roads, drinking water supply schemes, community welfare centers, facilities/furniture/lab equipment, etc. For schools, play ground, sanitation facilities, street lighting, agricultural land, horticultural camps and facilities.
- *Infrastructural facilities:* The Nigam will build such infrastructural development works in the vicinity of the project area that may be essentially required then construction of the project and or benefit the local population. These works may be mutually decided with Government of Himachal Pradesh.

First Plan is rehabilitation and resettlement plan (R&R plan) whose provisions is to provide land to those people who become landless after their land acquisition, construct houses, and to give cash for the house acquired. To provide employment to one family member of landless project affected families, allot shop in shopping complex, assistance for physical mobilization for displaced families. The Second Plan is Remedial Action Plan (REAP) whose provisions are the introduction of mobile health van, development of basic amenities in affected villages, income generation scheme for such families.

3.4.3 Resettlement of Landless Families

Approximately 224 hectares of private land has been acquired from 480 families for the NJPC project. This owners of the land acquired have been compensated for their land as per the rated fixed by the Government of Himachal Pradesh. Those families who left with more than 5 Bighas land after acquisition of part of their land were paid cash compensation only. However, those families, which were rendered landless their remaining land being less than 5 Bighas, have been provided alternative developed land by NJPC. Under the Antodaya programme of poverty alleviation, the Government of H.P. implemented a scheme wherein those families who owned landless than 5 Bighas (one acre) were provided additional land from out of the village common land so as not to live any rural family with less than 5 Bighas land. The NJPC has allotted developed agricultural land to each family who is rendered landless after land acquisition, equivalent of the area acquired or 5 Bighas whichever is less. The average area of land allotted per landless PAFS by NJPC comes to 2.2. Bighas (0-.175 ha.). Forty-one families from Jhakri and 21 from Kotla village were rendered landless. Out of 41 landless PAFs from Jhakri 37 families have been provided alternative developed land in Jhakri itself. In Kotla village initially 21 PAF'S were identified as landless who were to be allotted land at Nogli village by the Govt. of H.P. The PAF's did not agree to take either land earmarked for allotment for them at Nogli because of its very poor quality soil and its distance from their original village. An expert team from H.P. Agricultural University, Palampur on the request of NJPC conducted soil test of this Nogli village land. The findings of the soils test confirmed that the apprehension of the PAF's of Kotla were correct. The NJPC requested the Government of H.P. to explore the availability of alternative land for the distribution to landless PAF's of Kotla village. However, the latest situation is that as per the verification by SDM, Rampur out of 21 landless families of Kotla only 4 are eligible and all the 4 landless PAF's for Kotla have been provided alternative developed land in Kotla village itself.

3.4.4 Resettlement of Houseless Families

Each displaced family which has been rendered houseless on account of acquisition of house land for the project has been provided house within a built up plinth area of 45 square meter or alternatively has been paid cash grant of Rs. 45,000 to construct house with at least 45 square meter plinth area. If the plinth area was less than 45 square meters the cash grant was reduced accordingly. Out of the 61 families, who were rendered houseless by the project 43 opted for cash compensation, which has been paid to them. And 18 opted for house they were given alternative constructed house in resettlement colony at Jhakri. The physical mobilization of the oustee families to the new houses was done at the project cost. The water supply, electricity, street and approach paths in the rehabilitation colony have been provided at the project cost. Majority of the villages house in the project area were temporary Kucha structure (mud wall with tin or slate roofs). Those PAF's who were given cash compensation had now built permanent and bigger houses. Similarly, the alternative houses provided by the NJPC in the resettlement colony are of permanent nature. The PAF's feel that compensation grants in lieu of old house were sufficient to build a new permanent house. Some households have built bigger houses with compensation money plug some money from their own source.

3.4.5 Resettlement of Shopkeepers

A total of 87 shopkeepers have been displaced by the project. Out of these 79 were eligible for shop plots in market complex developed by the NJPC at Jhakri. Till date 71 shopkeepers have been allotted developed plots and remaining 8 are yet to get plots as the land development is in progress. Out of 71 displaced shopkeepers who were provided alternative shop plots in the market complex, 34 have constructed their shops and 14 of them have already started their business from the new shops. The NJPC has provided water supply, sewerage system, streetlight and other amenities in the market complex. Since the shop plots provided are a little away from the main highway and old market the trading has not yet fully shifted to the market complex. Therefore, some displaced shopkeepers who have been allotted shop plots have also taken shops on rent in the old market and continue to operate from the remaining old market at Jhakri. Nevertheless, they have also taken possession of plots/shops in the new market complex where they would shift when the new market wills becomes fully operational. All shops in the new market complex are of permanent structure unlike old shops, which were mostly kucha structure.

3.4.6 Employment to Landless Families

In the rehabilitation and resettlement plan of the NJPC

there is a provision the NJPC would provide employment to one member each of the landless PAF's. The SDM Rampur has done the identification of landless families who are affected by the project. Out of 62 families who are rendered landless by the project one person each from 51 families has already been provided a regular employment in the NJPC according to his or her capability and qualification. To provide widespread defused benefits to more families and to avoid multiple benefits accruing to same family, the NJPC ensured that those PAF's who have been provided employment shall not be eligible for allotment of shops in the market complex constructed by the NJPC at Jhakri and *vice versa.* Among 51 persons who were provided employment in the NJPC, 29 percent were women. Since the qualification of the candidate was below 12 the standard, without any vocational training therefore, the jobs provided to them are of unskilled nature mainly as attendants. However, some of them who acquired skills of computers word-processing while in employment at the NJPC have been promoted as clerks. Because some landless PAF's failed to nominate eligible members for employment in the NJPC, the full target of providing employment to 62 persons from landless PAF's could not be achieved. Some PAF's who are having more than one unemployed persons in the family could not resolve as to which members from the family should be nominated for employment. Project authorities also considered proposals for award of petty contract to the co-operative societies formed by PAF's on preferential basis so that some of them could be engaged in such jobs.

3.4.7 Resettlement of Families who got Cash Compensation Only

The NJPC paid compensation to PAF's for the land acquired according to the compensation rates fixed by the Rural Development Department for various types of lands. Out of this cash compensation some of the PAF's have purchased vehicles for commercial purposes and thus acquired an asset, which yields regular flow of income and employment to the family. Some of the households have constructed permanent houses out of the land compensation money received by them and have rented out the new house, earning regular monthly income for the family. Some households have put their compensation money in the term deposit accounts in the banks and are earning interest on it. However, there are some households who have used the cash compensation of their land acquired by NJUPC on non-productive expenditures marriages of their sons and daughters, treatment of diseases, purchase of household goods, etc.

3.4.8 Grants Provided Under Income Generation Scheme

The NJPC has started an income generation scheme to assist project-affected persons. Under the income generation scheme NJPC some unemployed members of PAF's are encouraged to take up non-land-based income generation activities, such as weaving, knitting, bee-keeping, tailoring, grocery shop, small dairy, etc. for diversifying their household incomes. Under this scheme NJPC provides a financial grant of Rs. 15,000 per family for particular activity. Only 33 PAF's availed assistance from NJPC under this scheme. The income generation schemes of NJPC did not achieve desired success (Bhatti, Singh, & Vaidya, 2002). Because the response for availing the benefits of this scheme was poor, second, the rate of diversion of grant to other activities than the required ones was very high. The survey revealed that on an average person running daily need shops under this scheme earns a net profit of Rs. 60 per day from the initial investment of about 11,000. Those people who started rearing crossbred cows under this scheme after meeting their cost are earning net income Rs. 70 per day. The development of NJPC colony at Jhakri village has created a sizeable market for milk where farmers are selling milk @ Rs. 12 per liter as compared to Rs. 9 per liter. It was suggested by the villagers that while providing financial assistance for self-employment income generation, the technical and marketing know how should also have been provided to them. The lack of previous experience has resulted in failures and low profits in the new ventures started by PAF's with financial assistance from the income generation scheme of the NJPC.

3.4.9 Mobile Health Unit and Hospital

The Mobile Health Unit (MHU) of the NUJPC started functioning in January 2000. The unit has a team consisting of a male medical doctor, a pharmacist and a driver, who tour villages in the van, which is fitted with medical equipments. The team makes four visits per week; two in project affected villages of Kinnaur district and two of the Shimla district. About 50 patients are examined daily in Shimla and 80 patients per day in Kinnaur. More than twelve thousand patients have been examined and treated by the MHU so far. The common ailment are diagnosed in the van itself and the patients are provided medicines there itself. However, the chronic patients are referred to the hospital for further thorough examinations. In general the village people are not fully aware about their various health problems. But as a result of visits by MHU, the awareness of the villagers about basic

3.5 DOMINO-EFFECT OF RESETTLEMENT REHABILITATION POLICIES

In the year 2002, a study was commissioned by Nathpa Jhakri Power Commission (NJPC) to assess the results of the Resettlement Action Plan implementation and to assess the impacts in terms of changes in the living standards of the project affected persons in terms of income, occupation, consumption pattern, housing standards, assets and land ownership, and by improving basic amenities in the affected villages, etc. In this study the magnitudes of indices during 2002 (after the programme implementation) are compared with the base line data (1996 situation) and with the control sample household data (households in the project area which are not affected by the NJPC project). The data reveal that the family size of the Project Affected Families (PAFs) has declined from 7.14 to 5.44 persons per family. The sex ratio has declined considerably (from 893 to 850 females per 1000 males). The proportion of minor individuals in the family accounted for 36 percent in 1996, which is now 27 percent. The proportion of old persons declined from 6 percent to 3 percent. Literacy rate has increased from 58 percent to 73 percent. The average annual household income (at 1996 prices) during the base line period was

Rs. 21,648 while in 2002 it has gone up to Rs. 76,575. At current (2002) prices the base line income comes to Rs. 29,114 and 2002 income was Rs. 1,04,640. In 2002 income of control households was Rs. 67,596. Hence, it is quite clear that after rehabilitation the income of PAFs has improved considerably when compared with base line income or control household income. Proportion of families living below poverty line has decreased from 25.6 percent to 16.8 percent. The average per capita monthly expenditure of PAFs has increased from Rs. 575 to Rs. 674, showing an improvement in their consumption pattern and standard of living. The percentage of workers engaged in regular employment has increased from 20 per cent to 30 per cent, while in agriculture it has declined from 72 per cent to 61 per cent. There has been a slight increase in the proportion of workers engaged in business activities (i.e. from 7 per cent to 9 per cent). A significant change has been noticed in the housing situation. Now more people live in *pucca* houses (45 per cent as compared to 11 per cent earlier), more families have now separate bathrooms (46 per cent as compared to 21 per cent earlier), and have separate toilets within house (39 per cent as against 16 per cent earlier). However, the average size of land-holding of the families has declined from 1.21 hectare to 0.373 hectare. As compared to 1996 data the overall yield rate of maize has not changed but that of wheat has increased. Number of all types of livestock owned by the PAFs has declined; the major decline was in sheep population. Before 1996 on an average a family owned 7.7 sheep, the number of which declined to 0.7 only in 2002 which was due to the effects of NJPC project. Since holding sizes declined, the requirement of draught animals also declined. The study concluded that the overall living standard of the families has improved due to NJPC project implementation. There is significant increase in the proportion of workers in the regular employment, especially with NJPC and with its contractors as daily wage earners. The earning capacity of those who were below poverty line has increased and thus some of them have crossed over to above poverty line families. Diversification of income and employment avenues through income generation schemes, towards business and other self-employment activities is taking place. There is improvement in the housing standard. The quality of health care has also improved due to enhancement of diagnostic facilities with the introduction of mobile health unit by MHU, which tours villages in the project area. The NJPC has taken measures to strengthen the existing infrastructure facilities, including health facilities and education and roads which are providing immense benefits to the PAFs in the project area. Project affected families have received full and adequate compensation. The compensation amount has been used rationally and judiciously by the families. Overall situation of all the project affected people is better now.

3.6 GENERAL FEATURES OF THE SJVN PROJECT AFFECTED VILLAGES

General features of the selected project affected villages, viz. Koyal, Boyal, Dutt Nagar, Tunan, Poshna and Nirmand have been discussed in this section.

3.6.1 Socio-Economic Infrastructural Facilities

Economic infrastructure, i.e. roads, transport, communication and power, have a significant role in promoting economic development by creating a base on which a higher level of economic activity can be carried out. Infrastructure represents if not the 'engine', then the 'wheels' of economic activity. Similarly, investment in social infrastructure like health and education constitutes the core of the economics of human resources. It is the quality of the people in terms of their physical and mental ability that ultimately determines the success of developmental policies.

Table 3.1 depicts the various socio-economic infrastructural facilities in the project affected villages. It has been analyzed whether such facilities are present in the village and if not the nearest location of such facilities and the distance involved. Provision of roads is one of the important modes of transportation, especially in hilly area. For example, village Dutt Nagar is linked with *pucca* road whereas village Koyal is linked with *kuccha* road. Remaining villages are linked with kuccha pedestrian path. Primary schools are present in all the villages but secondary schools are not. Similarly, the facilities of bank, post office and telephone are not available in any of the villages.

TABLE 3.1
Socio-Economic Infrastructural Facilities in Selected Villages

Social and Economic Infrastructural Facilities	*Koyal*		*Bayal*		*Dutt Nagar*	
	Location	*Distance*	*Location*	*Distance*	*Location*	*Distance*
(1)	*(2)*	*(3)*	*(4)*	*(5)*	*(6)*	*(7)*
Pucca Road	*Kuccha*	0	Up Avary Kaccha	4 km.	*Pucca*	With in Village
Nearest Railway Station	Shimla	130 km.	Shimla	126 km.	Shimla	117 km.
Bus Stand	Koyal	0	Bayal	0	Datt Nagar	0
Market	Rampur	14 km.	Rampur	12 km.	Rampur	13 km.
Primary School	Koyal	Within Village	Bayal	0	Datt Nagar	0
Secondary School	Datt Nagar	18 km.	Rampur	12 km.	Datt Nagar	0
Bank	Bayal, Kangra Cooperative Bank	4 km.	Bayal, Kangra Cooperative Bank	0	Nirath	10 km.
Post Office	Koyal	Within Village	Koyal	4 km.	Datt Nagar	0
Telephone	Nirmand	15 km.	Nirmand	11 km.	Rampur	13 km
Hospital	Rampur	15 km.	Nirmand	11 km.	Rampur	15 km

Primary Health Centre	Nirmand	15 km.	Nirmand	11 km	Nirath	0
Veterinary Hospital	Nirmand	15 km.	Koyal	3 km.	Nersu	3 km.
Veterinary Dispensary	Koyal	Within Village	Koyal	3 km.	Nersu	3 km.
Panchayat Centre	Bayal	4 km.	Bayal	0	Dutt Nagar	0
Electrification Status	Yes	—	Yes	—	Yes	0
Village Drainage	Yes (*pucca*)	—	No	—	Yes (*pucca*)	0
Provision of Sewerage	No	—	No	—	No	0
Sources of Drinking water	IPH Scheme	10 km.	Yes	0	Tap Water	0

(Contd.)

TABLE 3.1 (Contd.)

Social and Economic Infrastructural Facilities	*Tunan*		*Poshna*		*Nirmand*	
	Location	*Distance*	*Location*	*Distance*	*Location*	*Distance*
(1)	*(8)*	*(9)*	*(10)*	*(11)*	*(12)*	*(13)*
Pucca Road	*Pucca*	12 km.	Chuna Ghai	4 km.	—	0
Nearest Railway Station	Shimla	155 km.	Shimla	150 km.	Shimla	137 km.
Bus Stand	Rampur	12 km.	Rampur	5 km.	Nirmand	5 km.
Market	Rampur	12 km.	Rampur	5 km.	Nirmand	0
Primary School	Tunan	0	Up to 3rd alternative school running by SSA	With in Village	Nirmand	0
Secondary School	Rampur Arsu	12 km. 14 km.	Middle school Kisholi	1 km.	Nirmand	0
Bank	Grameen bank	12 km.	Jagat Khana	4 km.	PNB, Cooperative bank	0
Post Office	Tunan	0	Poshna	0	Nirmand	0
Telephone	No	—	Jagat Khana	0	Nirmand	0
Hospital	Rampur	16 km.	Rampur, Nirmand	15 km. 22 km.	Nirmand	0
Primary Health Centre	Rampur	16 km.	Kisholi	1 km.	Nirmand	0

Veterinary Hospital	Rampur	12 km.	Kisholi	1 km.	Nirmand	0
Veterinary Dispensary	Rampur	12 km.	Kisholi	1 km.	Nirmand	0
Panchayat Centre	Jagat Khana	12 km.	Brow	5 km.	Nirmand	0
Electrification Status	Yes	0	Yes	0	Yes	0
Village Drainage	No	0	Yes (*Kutcha*)	0	No	0
Provision of Sewerage	No	0	No	0	No	0
Sources of Drinking water	Kandlu Nala	1 km.	Spring	100 Meter	Nirmand	0

Source : Department of Census Operations, Shimla.

Most of the health-related facilities are available only in village Nirmand.

3.6.2 Socio-Economic Characteristics

The socio-economic characteristics of selected villages based on 2001 census data have been given in Table 3.2. The analysis reveals that village Nirmand is the largest village having a total population of 6200 and village Boyal, the smallest with a total population of 636 only. It is seen in the Table that the ratio of male and female population in these villages is quite satisfactory. Highest literacy percentage of about 82 percent was found in village Poshna whereas village Tunam was at the bottom in this respect, with the literacy percentage of about 62 per cent.

3.6.3 Occupational Structure

Table 3.3 presents the occupational structure of project affected villages. The data in the Table reveals that highest number of workers is in village Nirmand. The details of sex-wise main workers, cultivators, marginal workers and non-workers have also been presented in the Table. The highest numbers of these are invariably found in Nirmand and smallest in Boyal, which indeed are determined by the size of population.

3.6.4 Operational Holdings in Selected Villages

The details of operational holdings in selected villages, based on village revenue records, have been presented in Table 3.4 indicating that the highest number of operational holdings, 819 covering an area of about 631 hectares is in Nirmand. This is followed by Tunan having 771 holdings spread over an area of 515 hectares and Poshina heaving 451 holdings covering an area of about 250 hectares. The lowest number of operational holdings is in Boyal with 87 holdings in an area of 43 hectares. The size-wise details about number and area of operational holdings can also be seen in this Table.

3.6.5 Land Utilization Pattern

As per village records, Village Nirmand is the largest village having a total geographical area, as per village revenue

TABLE 3.2
Socio-Economic Characteristics of Selected Villages (2001)

Features	*Villages*					
	Koyal	*Bayal*	*Dutt Nagar*	*Tunan*	*Poshna*	*Nirmand*
(1)	*(2)*	*(3)*	*(4)*	*(5)*	*(6)*	*(7)*
No. of Households	484	114	250	824	773	1380
Total population	2471	636	1213	4115	3522	6200
Male	1174	265	540	1756	1630	2864
Female	1037	284	542	1689	471	2616
Children	260	87	131	670	421	720
Literacy %	65.35	66.48	77.26	61.68	81.61	74.16
Male	74.87	78.86	88.33	75.51	89.14	85.64
Female	54.58	54.92	66.23	47.30	73.28	61.58
SC Population	743	299	519	2303	1368	2953
ST Population	1	5	14	7	63	32
SC & ST Population	744	304	533	2310	1331	2985
IRDP families	—	26	—	112	—	120
No. of households under IAY	—	10	—	2	—	2
No. of households under SJRY	—	30	—	4	—	—

Source : Department of Census Operations, Shimla.

TABLE 3.3
Occupational Structure of Selected Villages (2001)

Occupation	*Koyal*	*Bayal*	*Dutt Nagar*	*Tunan*	*Poshna*	*Nirmand*
Total workers	1638	318	675	2148	1564	3436
Male	858	147	339	1095	907	1805
Female	780	171	336	1053	657	1631
Main workers	1289	312	659	1984	1321	2848
Male	667	141	329	1038	853	1643
Female	622	171	330	946	468	1205
Cultivator	1184	300	562	1548	681	1955
Male	574	131	252	625	380	856
Female	610	169	310	923	301	1099
Main other workers	105	12	97	436	640	893
Male	93	10	77	413	473	787
Female	12	2	20	23	167	106
Marginal workers	349	6	16	164	243	588
Male	191	6	10	57	54	162
Female	158	0	6	107	189	426
Non-workers	833	318	535	1967	1958	2764
Male	457	168	278	1004	941	1423
Female	376	150	260	963	1017	1341

Source : Department of Census Operations, Shimla.

TABLE 3.4
Operational Holdings in Selected Villages

(Area in Hectares)

Size of holdings	*Koyal*		*Bayal*		*Dutt Nagar*		*Tunan*		*Poshna*		*Nirmand*	
	No.	*Area*	*No.*	*Area*	*No.*	*Area*	*No.*	*Area*	*No.*	*Area*	*No.*	*Area*
(1)	*(2)*	*(3)*	*(4)*	*(5)*	*(6)*	*(7)*	*(8)*	*(9)*	*(10)*	*(11)*	*(12)*	*(13)*
Marginal	145	40	75	23.91	184	78	616	253	371	97.91	568	144.12
Small	11	15	9	11.85	49	65	128	183	42	45.08	208	258.26
Medium	5	22	3	7.24	24	39	23	60	33	87.25	37	97.80
Large	1	8	—	—	8	53	4	18	5	26.93	6	31.06
Total	162	85	87	43.0	265	236	771	515	451	257.71	819	631.24

Source : Village revenue records.

papers, of 1102 hectares and this is followed by Dutt Nagar and Tunan. Village Boyal has the smallest geographical area, 76 hectares. There was low area under forests in all the villages. Except for village Koyal there was some are a 'not available for cultivation' and Nirmand topped in this respect. There was no area under non-agricultral uses in Boyal and Poshna. Barren and uncultivable land was present only in Koyal, Boyal and Dutt Nagar. Dutt Nagar was the only village having land under permanent pastures and other grazing lands, and 539 hectares of land was under this category. Dutt Nagar also had 11 hectares of area under miscellaneous tree crops and groves. Fallow land was present only in Koyal and Tunan.

3.6.6 Cropping Pattern

According to village revenue records, maize and paddy are the most common kharif crops followed by black gram. Other crops grown in some of the villages include rajmash, potato, tomato, radish and kulth, etc. In rabi reason wheat is the most common crop grown in all the villages. In some of the villages barley, peas, gram, mustard, potato and cabbage, etc. are grown. All the villages have orchards of apple, almond, plum, etc.

3.6.7 Sources of Irrigation

Kuhls are the only source of irrigation in the selected villages, and each village has at least one khul according to village revenue records. Village Tunan was having three kuhls with the scope of irrigation of 6 hectares of agricultural land. On the other hand, Dutt Nagar was having two khuls but has a command area of 116 hectares. Koyal village has only one kuhl irrigating an area of 74 hectares.

Data regarding the general information in respect of project affected households, workers and non-workers, occupational pattern of workers, source of income, extent of income and families below poverty line have been appended as Appendices V to VIII, respectively. The main purpose of the above discussion was to highlight various dimensions of socio-economic development in Project Affected villages of Satluj River Valley.

3.7 HYDROPOWER PROJECTS ON THE SATLUJ RIVER VALLEY AND ECONOMIC DEVELOPMENT

The preceding section highlights the profile of Hydropower Projects located on the river Satluj and executed by Satluj Jal Vidyut Nigam Limited. However, evaluation of Economic Benefits of a particular power project is a challenging task; nevertheless an attempt has been made to evaluate the contribution of Hydropower Projects of SJVN Limited to Economic Development of Himachal Pradesh in the present section.

3.7.1 Socio-Economic Development Status in Respect of PAF *vis-à-vis* Districts of Shimla, Kinnaur and State average

SJVN Hydropower Projects, being executed by a public sector corporation, i.e Satluj Jal Vidyut Nigam Limited, has compensated the project affected families judiciously and rationally. Consequently, the socio-economic conditions of project affected families have improved remarkably. An endevour has been made to verify this fact with the help of some standard indicators of social and economic development as presented in Table 3.5.

Sex Ratio (which indicates number of females per thousand of males) is an important component of social development. It is one of the significant determinants of birth rate and death rate as also an indicator of gender discrimination. It determines the marriage rate and number of children per couple. Less female ratio leads to many social and moral evils. Hamirpur district had the highest sex ratio (1099) according to 2001 census followed by Kangra (1025) and Mandi (1012) districts of Himachal Pradesh. Since, SJVN Hydropower Projects are mainly located in sub-areas of districts of Shimla and Kinnaur, the comparison of project area with the average of both these districts emphasizes the authentic stipulation of the project affected area. In the matter of sex ratio, project affected area was better placed not only in comparison to Shimla and Kinnaur districts but also in respect of the state average.

TABLE 3.5

Socio-Economic Status of Project Affected Families vis-à-vis Shimla, Kinnaur Districts and the State Average

Sl. No.	Indicators	*Project Affected Area	Shimla#	Kinnaur#	Himachal Pradesh#
1.	Sex ratio	1037	896	857	968
2.	Literacy (%)	83.10	79.1	75.2	76.5
3.	Female Literacy (%)	76.0	70.1	64.4	67.4
4.	Literacy (%) among SC/ST population	77.00	55.66	52.25	50.14
5.	Female Literacy (%) among SC/ST Female population	68.00	43.85	37.81	36.1
6.	Main workers to total population (%)	60.37	37.27	49.92	32.32
7.	Female workers to total female population (%)	61.45	38.29	42.18	21.07
8.	Families below poverty line (%)	0.00	33.67	26.57	27.62

Data compiled are according to 2001 census.

Sources : *Base line survey data (Survey conducted in April 2005) are taken from the Baseline Demographic Socio-Economic Survey Satluj Jal Vidyut Nigam Ltd., Shimla, September 2006.

A person aged seven and above, who can both read and write with understanding in any language is treated as literate. Literacy is considered yet another component of social development. According to the census of 2001, among the districts of Himachal Pradesh, Hamirpur was at the top in respect of both overall literacy rate and female literacy rate with value of literacy percentage being 83.16 and 76.41 respectively. It is evident from Table 3.5 that the values of overall literacy percentage and female literacy percentage of Nathpa Jhakri Project affected area are not only higher than that of Shimla and Kinnaur districts and the State average but closely approaching the value of Hamirpur district.

Poverty ratio and employment ratio are important components of economic development. Table 3.5 clearly reveals

FIG. 3.1
Socio-Economic Status of Project Affected Families vis-à-vis Shimla, Kinnaur and the State Average

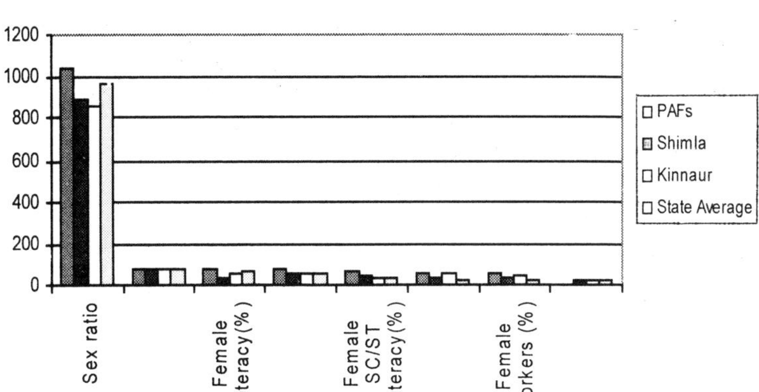

that the SJVN project affected area is better placed in respect of both these components of economic development.

3.7.2 Measurement of Level of Development

To begin with, the level of development in terms of each indicator (as given in Table 3.5) was worked out. The level was seen in relation to the state average. The SJVN Project Area is treated as one separate unit just like districts with the intention of ascertaining the level of development. All the districts along with SJVN Project Area were arranged in the descending order of sex ratio, percentage of overall literacy, percentage of female literacy, literacy among overall and SC/ST population, percentage of main workers to total population, percentage of female workers to total female population and families below the poverty line.

In Table 3.6 the level of development has been shown. The level of all those districts and Project Affected Area is termed as 'high' if it exceeds the state average. On the other hand, the level of districts below the state average is labelled 'low'. This procedure is followed for all the indicators.

It may be observed from Table 3.6 that SJVN Project Affected Area is invariably above the state average in almost all the indicators of socio-economic development. It is pertinent to

Table 3.6
Level of Socio-Economic Development

Sl. No.	Indicators	High Level of Development	Low Level of Development
	(1)	(2)	(3)
1.	Sex ratio (Number of females per thousand of male population)	1. Hamirpur	1. L&S
		2. SJVN Project Affected Area	2. Solan
		3. Kangra	3. Kinnaur
		4. Mandi	4. Shimla
		5. Una	5. Sirmaur
		6. Bilaspur	6. Kullu
			7. Chamba
2.	Literacy Percentage	1. Hamirpur	1. Chamba
		2. SJVN Project Affected Area	2. Sirmaur
		3. Una	3. L&S
		4. Kangra	4. Kullu
		5. Shimla	5. Kinnaur
		6. Bilaspur	
		7. Solan	
3.	Female Literacy (%)	1. SJVN Project Affected Area	1. Chamba
		2. Hamirpur	2. Sirmaur
		3. Una	3. L&S
		4. Kangra	4. Kullu
		5. Shimla	5. Kinnaur
		6. Bilaspur	6. Mandi
		7. Solan	
4.	Literacy percentage among SC/ST population	1. Hamirpur	1. Chamba
		2. Una	2. Sirmaur
		3. Kangra	
		4. SJVN Project Affected Area	
		5. Shimla	
		6. Solan	

TABLE 3.6 (Contd.)

(1)	(2)	(3)
	7. Kullu	
	8. Mandi	
	9. L&S	
	10. Bilaspur	
	11. Kinnaur	
5. Female Literacy (%) among SC/ST Female population	1. SJVN Project Affected Area	1. Chamba
	2. Hamirpur	2. Sirmaur
	3. Una	3. L&S
	4. Kangra	
	5. Shimla	
	6. Kullu	
	7. Bilaspur	
	8. Solan	
	9. Mandi	
	10. Kinnaur	
6. Percentage of main workers to total population	1. SJVN Project Affected Area	1. Chamba
	2. L&S	2. Hamirpur
	3. Kinnaur	3. Kangra
	4. Kullu	4. Una
	5. Sirmaur	5. Mandi
	6. Shimla	
	7. Solan	
	8. Bilaspur	
7. Percentage of female main workers to total female population	1. SJVN Project Affected Area	1. Una
	2. L&S	2. Kangra
	3. Kinnaur	3. Chamba
	4. Kullu	4. Solan
	5. Shimla	
	6. Sirmaur	
	7. Bilaspur	

(*Contd.*)

TABLE 3.6 (*Contd.*)

(1)	*(2)*	*(3)*
	8. Mandi	
	9. Hamirpur	
8. Percentage of Families below poverty line (Poverty ratio)	1. SJVN Project Affected Area	1. Chamba
	2. Una	2. L&S
	3. Kullu	3. Shimla
	4. Sirmaur	
	5. Kangra	
	6. Hamirpur	
	7. Mandi	
	8. Kinnaur	
	9. Bilaspur	
	10. Solan	

mention here that the ranking of SJVN project affected area is also at forefront which shows that the efforts made by Nathpa Jhakri Corporation for the socio-economic development of the project affected area are quite commendable.

3.7.3 Measurement of Development Index

United Nations Institute of Social Research (UNISR: 1991) uses the following method for measuring social development. The index is derived as follows:

$$\text{Deprivation Score} = \frac{\text{Maximum Value} - \text{Actual Value}}{\text{Maximum Value} - \text{Minimum Value}} \quad \ldots (1)$$

The deprivation score ranges between 1, denoting maximum deprivation, and 0 representing least deprivation. Subtracting the deprivation score from 1, the development index is worked out as:

$$\text{Development Index} = 1\text{—Deprivation Score} \quad \ldots (2)$$

The development index ranges between 1, denoting maximum development, and 0 representing no development. It measures the relative development level of various regions in relation to the most developed one. This measures the extent to which a particular region is lagging behind as compared to the one at the top. The UNISR, however, makes use of a simpler formula. The development score is calculated directly by ascertaining the position of each region *vis-à-vis* the most backward ones. The results, however, are virtually the same by both the methods.

This method has been used for estimating the development index of the SJVN Project affected area as well as the districts of Himachal Pradesh in respect of the various indicators of socio-economic development in the Tables 3.7 to 3.15 below.

Table 3.7
Development Index in Respect of Sex Ratio

Sl. No.	*District/Region*	*Development Index*
1.	SJVN Project Affected Area	0.77
2.	Bilaspur	0.60
3.	Chamba	0.48
4.	Hamirpur	1.00
5.	Kangra	0.73
6.	Kinnaur	0.10
7.	Kullu	0.36
8.	L&S	0.00
9.	Mandi	0.68
10.	Shimla	0.24
11.	Sirmaur	0.36
12.	Solan	0.08
13.	Una	0.62

Table 3.8
Development Index in Respect of Literacy Percentage

Sl. No.	*District/Region*	*Development Index*
1.	SJVN Project Affected Area	0.99
2.	Bilaspur	0.78
3.	Chamba	0.00
4.	Hamirpur	1.00
5.	Kangra	0.88
6.	Kinnaur	0.60
7.	Kullu	0.50
8.	L&S	0.50
9.	Mandi	0.63
10.	Shimla	0.83
11.	Sirmaur	0.37
12.	Solan	0.70
13.	Una	0.90

Table 3.9
Development Index in Respect of Female Literacy Percentage

Sl. No.	*District/Region*	*Development Index*
1.	SJVN Project Affected Area	1.00
2.	Bilaspur	0.76
3.	Chamba	0.00
4.	Hamirpur	0.99
5.	Kangra	0.89
6.	Kinnaur	0.57
7.	Kullu	0.44
8.	L&S	0.44
9.	Mandi	0.59
10.	Shimla	0.78
11.	Sirmaur	0.42
12.	Solan	0.66
13.	Una	0.89

TABLE 3.10

Development Index in Respect of Literacy Percentage among SC/ST Population

Sl. No.	District/Region	Development Index
1.	SJVN Project Affected Area	0.88
2.	Bilaspur	0.37
3.	Chamba	0.00
4.	Hamirpur	1.00
5.	Kangra	0.72
6.	Kinnaur	0.34
7.	Kullu	0.41
8.	L&S	0.40
9.	Mandi	0.40
10.	Shimla	0.41
11.	Sirmaur	0.01
12.	Solan	0.41
13.	Una	0.88

TABLE 3.11

Development Index in Respect of Female SC/ST Literacy among Female SC/ST Population

Sl. No.	District/Region	Development Index
1.	SJVN Project Affected Area	1.00
2.	Bilaspur	0.43
3.	Chamba	0.02
4.	Hamirpur	0.95
5.	Kangra	0.52
6.	Kinnaur	0.36
7.	Kullu	0.44
8.	L&S	0.32
9.	Mandi	0.38
10.	Shimla	0.49
11.	Sirmaur	0.00
12.	Solan	0.42
13.	Una	0.64

TABLE 3.12
Development Index in Respect of Percentage of Main Workers to Total Population

Sl. No.	District/Region	Development Index
1.	SJVN Project Affected Area	0.84
2.	Bilaspur	0.25
3.	Chamba	0.31
4.	Hamirpur	0.29
5.	Kangra	0.00
6.	Kinnaur	0.87
7.	Kullu	0.65
8.	L&S	1.00
9.	Mandi	0.33
10.	Shimla	0.37
11.	Sirmaur	0.27
12.	Solan	0.44
13.	Una	0.05

TABLE 3.13
Development Index in Respect of Female Main Workers to Total Female Population

Sl. No.	District/Region	Development Index
1.	SJVN Project Affected Area	1.00
2.	Bilaspur	0.09
3.	Chamba	0.00
4.	Hamirpur	0.29
5.	Kangra	0.11
6.	Kinnaur	0.64
7.	Kullu	0.53
8.	L&S	0.79
9.	Mandi	0.30
10.	Shimla	0.36
11.	Sirmaur	0.34
12.	Solan	0.14
13.	Una	0.07

TABLE 3.14
Development Index in Respect of Poverty Ratio*

Sl. No.	*District/Region*	*Development Index*
1.	SJVN Project Affected Area	1.00
2.	Bilaspur	0.57
3.	Chamba	0.00
4.	Hamirpur	0.61
5.	Kangra	0.61
6.	Kinnaur	0.56
7.	Kullu	0.69
8.	L&S	0.38
9.	Mandi	0.60
10.	Shimla	0.45
11.	Sirmaur	0.61
12.	Solan	0.55
13.	Una	0.69

*Reverse value has been calculated for ascertaining the development index.

The above Tables 3.7 to 3.14 indicate development index of different districts along with Nathpa Jhakri Project Affected Area. In order to see the composite picture of development, a composite development index has also been worked out which is presented in Table 3.15. It is interesting to notice in the Tables above that the SJVN Project affected area enjoy a development index score in respect of all the indicators higher than that of all the districts.

Table 3.15 clearly reveals that in the matter of overall socio-economic development, the SJVN Project Affected Area was at the top followed by Hamirpur. The very low level of development was found in district Chamba. Hence, SJVNL has made sufficient effort to improve the quality of life in the project affected area.

Socio-economic aspects of population, especially those who reside near the project site, are one of the most important considerations for any development project. The construction phase of a project generally has a pronounced impact on the

TABLE 3.15
Development Index in Respect of Overall Socio-Economic Development

Sl. No.	*District/Region*	*Development Index*
1.	SJVN Project Affected Area	0.94
2.	Bilaspur	0.48
3.	Chamba	0.11
4.	Hamirpur	0.77
5.	Kangra	0.56
6.	Kinnaur	0.51
7.	Kullu	0.51
8.	L&S	0.48
9.	Mandi	0.49
10.	Shimla	0.49
11.	Sirmaur	0.30
12.	Solan	0.43
13.	Una	0.59

FIG. 3.2
Development Index in Respect of Overall Socio-Economic Development

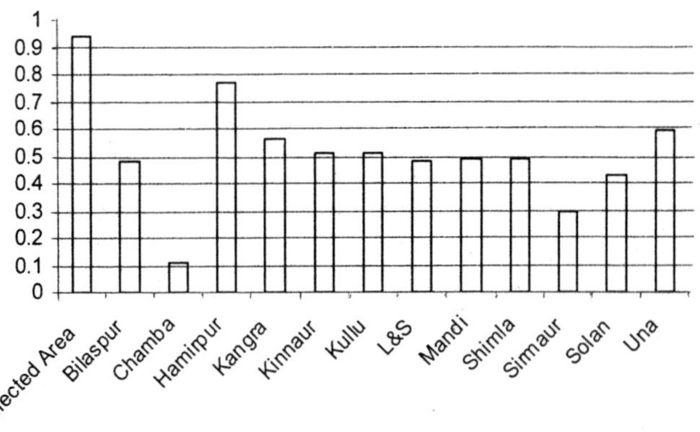

socio-economic aspects of the local population for the reasons that a section of the population faces displacement from their land which is acquired for the construction of the project and the construction activities bring in a sea change in the local environmental settings due to large scale influx of workers altering the traditional economic activities, etc. This area has witnessed inward migration of a large number of labour force including a drastic change in the socio-economic fabric of the local population. Therefore, it would be expected that majority of the impacts would be deleterious rather than beneficial, at the initial stage of the project.

SJVN has acquired approximately 224 hectare private land in districts Shimla and Kinnaur for the construction of NJHEP. All the land owners have been compensated for the land acquired as per the rates announced by the LAO under the Land Acquisition Act. To improve the living conditions and standards of the Project Affected Families (PAFs), SJVN has formulated and implemented a Rehabilitation and Resettlement (R&R) Plan for extending to them certain benefits in addition to the compensation paid for the acquired land. The salient features of this package are:

- Provision of developed agricultural land to landless PAFs.
- Provision of house or Rs. 45,000 according to their choice to each landless PAF whose house is acquired.
- Allotment of plots for shops at the Jhakri market complex for displaced shopkeepers.
- Preference in allotment of shops at the shopping complex at Jhakri to displaced shopkeepers.
- Provision of employment to one member of each displaced family according to his capability and qualification subject to availability of vacancy.

SJVN is extending these benefits to the landless families duly identified by the Sub-Divisional Magistrate, Rampur who is also R&R officer for this Project. Implementing the R&R Plan in letter and spirit, SJVN has extended these benefits to 62 landless families duly identified by the SDM, Rampur. Out of above families, 43 persons have been provided regular employment, 61 families were identified landless and houseless, out of whom 12 families have been allotted built up houses at R&R colony at Jhakri and 43 have been paid Rs. 45,000 each as per their option. Besides, 6 houses are under tendering stage for construction at Jhakri. These houses have a plinth area of 45 sq. m. each, and have been provided with electricity, water supply, sewerage system, street light upto this colony and 3.5 meters wide metalled and tarred road to approach the colony. The displaced shopkeepers are being allotted shop plots in the market complex at Jhakri on payment basis.

In addition to the above R&R benefits being given to landless, PAFs, SJVN has undertaken and completed many works for community development and area development. These include upgradation of various schools, construction of 27 kms Rampur bye-pass road, various link roads to villages, strengthening of balley bridges at Mangland, Nigulsari, Nogli and Sholding, construction of playground for schools, financial assistance of Rs. 8 crore for 200 beded hospital at Rampur and water supply schemes to various villages around the project.

Further, Nathpa-Jhakri Power Corporation has paid a dividend of Rs. 35.40 crores to Government of Himachal Pradesh for its operations for the year 2004-05. An interim dividend of Rs. 8.14 crores had already been paid by the SJVN to the State Government. From its nearly full year's commercial operations for the year 2004-05, the company has posted a profit of Rs. 298.43 crore on a turnover of Rs. 1098.28 crores. The company has declared a maiden dividend of Rs. 143.16 crore for the year 2004-05 to be shared by the equity partners Government of India and the Government of Himachal Pradesh in the ratio of 3:1 respectively. The Company's flagship venture 1500 MW Nathpa Jhakri Hydropower Project had commenced commercial operations in October 2003 and has since been supplying valuable power to the northern grid beneficiary states viz. Delhi, Rajasthan, Uttar Pradesh, Punjab, Haryana, Himachal Pradesh, Jammu & Kashmir and Chandigarh. The Satluj Jal Vidyut Nigam (formerly SJVN) is presently executing three more projects in Himachal Pradesh namely 1020 MW Khab Hydroelectric Project, 412 MW Rampur Hydroelectric Project and 700 MW Luhri Hydroelectric Project.

4

Environmental Concerns and Role of Hydropower Projects in Development

This chapter has been devoted to study the environment concerns of hydropower projects. Firstly, it briefly explains about the utilization of available natural resources for development. Secondly, it focuses on recent criticism of large Hydro projects on environmental considerations. Thirdly, it points out the problems of displacement and rehabilitation of project affected families. Fourthly, it highlights environment impact assessment, and finally, it briefly discusses the benefits of hydropower resources.

The economy of our country is largely based on agriculture. Many of the developing countries have an agricultural economic base that is in turn characterized by rather perennial food grain shortages. Frequent famines resulting from inadequate or ultimately rains are a common feature of these economies. The planners as well as the decision-makers, therefore, always stress the need for increased output of food grains through better seeds, fertilizers and intensive irrigation practices. Creation of a network of irrigation facilities is thus a priority area as far as the developmental efforts in the third world countries are concerned.

One common denominator of all the developing countries is their large population base. The situation is further aggravated by an alarming rate of increase of population that adds millions of new individuals to be fed, clothed and sheltered with the already limited resources. The increasing rate of increase in population is the large sink where every real benefit of several developmental activities submerges. This gives rise to two biggest polluters-hunger and poverty. Struggle for mere survival is unfortunately a stark reality in a majority of the developing countries.

4.1 UTILIZATION OF AVAILABLE NATURAL RESOURCES

The economic development of a country and rise in standard of living of people largely depends upon the development and utilization of available natural resources. Water, the major natural resource, holds the key to economic prosperity and stability of the nation. World-wide, Hydel power is the source to have been fully exploited by developed nations in order that they command a power generating source that is free from foreign control of sources of supply, and at the same time renewable. However the availability of water is not regular and its variability from season to season and year to year is very high. Large temporal and spatial variation in rainfall results in equally large variation in runoff of the rivers throughout the year. Consequently, run-of-the river schemes for irrigation and hydropower generation or small dams with small storages are not able to make available regular supply of water/power throughout the year. Optimum utilization of water for irrigation/hydropower generation is not possible unless its storage during the period of high availability is done by constructing large dams, where water during the monsoon season, when available in abundance, can be done, and subsequently utilized for augmenting the available lean period supplies for irrigation and generation of power.

River Valley Projects serve a basic necessity of a country whose economy is largely based on agriculture. The irrigation and power are crucial inputs for increased productivity. A high priority has, therefore, been given in the national planning process to the creation of the river valley projects ever since India became independent. The prominence given to the River Valley Projects in India's successive Five-Year Plans is consequent to the contribution to the prosperity of the country by the water resources projects.

4.2 RECENT CRITICISM OF LARGE HYDRO PROJECTS ON ENVIRONMENTAL CONSIDERATIONS

Concern for environmental pollution[1] is rather a recent phenomenon that emerged mainly from ill effect of industrial growth through the planning process that somehow overlooked the role of natural resources in developmental activities. Over the years, the information accumulated in course of working of river valley projects, revealed that the river valley projects, like all other development projects, have beneficial as well as adverse impacts. The two major criteria, that the project should maximize economic returns; and it should be technically feasible, are no longer considered adequate to decide the desirability or even the viability of the project. It is now widely recognized that the development effort may frequently produce not only sought for benefits, but either-often unanticipated-undesirable consequences as well, which may nullify the socio-economic benefits for which any project is basically designed. These impacts must be carefully assessed and balanced for achieving sustained benefits.

4.2.1 Main Areas of Criticism

The construction of large hydropower projects in India and elsewhere is being criticized mainly on the grounds that it involves:

- Displacement of people from their traditional homes due to submergence caused by large storages created by hydropower projects.
- Loss of traditional means of livelihood of the displaced people.

- Change in the flow regime in the river channels downstream of dams/Hydel projects.
- Environmental and ecological effects of large storages created by the dams/Hydel projects.

It is well known facts that the electricity cannot be generated in the vacuum. It requires acquisition of land or huge sites, native houses as well as agricultural and horticultural land that are submerged in the reservoirs. Sometime the movable and immovable properties are liable to disrupt by the formation of such huge reservoir. In hilly areas where the huge reservoir are not needed there the excavation of underground tunnel are practiced that when executed the internal harmony of the mountain are totally disturbed as well as hollowness and cracks appears in it and the houses of the native people. There also exists large deforestation during construction of the project for making site, road and colonies for the employees. Not only this, there exists various types of pollution in the surrounding area. Various types of dust, dirt, and formation of poisonous gases such as carbon dioxide, sulfur, nitrogen di-methyl ether, and chlorofluorocarbon, etc. smokes and heavy scale blasting have disrupted the lives of the human being. It surmounts pressure on the human being and on their fertile land by which the agricultural as well as horticultural crops are diminishing gradually and Pasteur for the animal is becoming scant, barren and infertile. Besides this, natural drinking water as well as cultivable water of rivers and streamlets. *Nadi* and *Nallas* continuously disrupting and water levels are going down.

The formation of Hydel projects are vanishing vegetation and restricting greenery of the environment. During rive side there use to be greenery that has dried by formation of reservoir for the projects. The harmony of interior region of the earth has disrupted that is inviting various types of the disaster. Now the area has become sensitive for seismic zone where earthquake as well as of fire due to its dryness can cause problem at any time. The garbage erupted from the project site has deleterious impact on the human and animal hygiene. Beside this undue rainfall, hot, spring and fluctuation in winter and summer seasons have been reported by the local people. By continuous pollution at various places it is causing green house effect and proving one of the factors in depleting ozone layer. The quantity of rainfall as well as snowfall has now been traced as scant. Sometime the area comes into the grip of flood as well as sliding problem are also frequently witnessing during rainy reason.[2]

4.3 DISPLACEMENT AND REHABILITATION PROBLEMS

The construction of hydropower projects, large or small, does involve displacement of people from their traditional homes that calls for advance planning in acquisition of land, evacuation and rehabilitation of the effected people.[2] The resettlement policy should be evolved on collaboration with the displaced people or their representatives by evolving a policy based on oustees' preference by:

- Advance planning in acquisition of land.
- Provision for giving land for land.
- Liberal compensation for submerged houses, land, trees, water mills, other affected properties.
- Provision of free plots and some agriculture land for landless.
- Provision of social and other amenities such as drinking water, schools, healthcare centers, recreational facilities in oustees' colonies.

It may also be noted that all type of developmental activities cause some type of displacement and other long-term effects and hydro projects are no exception. So far as the loss of traditional means of livelihood is concerned, large hydro-projects duly compensate the same by way of creating large means of employment opportunities. More comprehensive Resettlement and Rehabilitation policies are being implemented as prescribed by the Central government. The detail of the National Resettlement and Rehabilitation Policies are appended as Appendix-I in this book which has been framed by economic planners. Every care has been given in the policy so that the loss of homes, lands, etc. could be adequately substituted to the affected families with the provision of house for house, land for land as well as job opportunity including substantial monetary benefits for their properties.

4.4 ENVIRONMENT IMPACT ASSESSMENT

Implementation of policies and programmes relating to conservation of the country's natural resources including lakes and rivers, its biodiversity, forests and wildlife, ensuring the welfare of its animals and prevention and abatement of pollution are primary concerned of the Ministry of Environment and Forests. While implementing these policies and programmes, the Ministry is guided by the principle of sustainable development and enhancement of human well-being. For sustainable development and optimal use of natural resources, environmental considerations are required to be integrated in planning designing and implementation of development projects. Environmental Impact Assessment (EIA) is one of the proven management tools for incorporating environmental concerns in development process and also in improved decision-making. The programme of EIA, in vogue in the Ministry for the last two decades was initiated with the appraisal of River Valley Projects. To give legislative status to the procedure of impact assessment, EIA was made mandatory since January 1994 for thirty-two categories including River Valley and Hydroelectric projects of development activities.

It is very necessary for an agrarian country like ours to solve the problem of feeding the growing population and to build up its economy by development agriculture and industry on a sound footing to ensure availability of sustained power and abundant irrigation supplies. This can only be provided by large hydro projects. Environmental appraisal is an important responsibility and the function involves evaluation of environmental implications and incorporation of necessary safeguards for the activities having a bearing on environmental quality. The requirement is to consider environmental aspects as an integral part of a development project to achieve sustained development with minimum environmental degradation and prevention of long-term environmental side effects by incorporating mitigative measures. It is imperative to analyse whether the adoption of environmental measures is going to result in any short or long-term social and economic benefits or not. The object of environmental impact assessment is to ensure that development proceeds hand in hand with ecological preservation so as to achieve sustained growth. Environmental protection for sentimental reasons alone is neither desirable nor can it be defended.

4.4.1 Positive Impacts of Large Hydro Projects

Positive impacts of water resources projects outweigh the feared negative impacts and can be kept the minimum by integrating environmental concerns in the project planning right from the beginning. Large hydro projects have substantial positive environmental impacts, e.g. assured supply of drinking water and water for irrigation and industry, flood moderation, reduction in drought frequency, improved agricultural produce, production of hydropower, tourism, pisciculture, ground water recharge, development of industry and consequent economic progress and creation of additional employment opportunity, etc.

4.4.2 Environmental Impacts and Mitigation Measures in Hydropower Projects in Himachal Pradesh

Most of the projects under execution and projects planned for execution are run-of-the-river type development in Himachal Pradesh. Most of these projects involve a diversion structure in the form of a barrage or a trench weir, tunnel for water conductor system and underground power house. Thus, surface structures in many of the projects are mainly a barrage or weir and tail race outlet structure. In view of most of the projects being located underground, there is least environmental impact.

In order to minimize impact during construction, proper dumping areas with retaining structures and beams to reduce surface run-off along the excavates slopes, creation of setting structures for collection of waste oil, wetting of un-metalled roads regularly to prevent generation of dust, are strictly enforced. Maintenance of vehicles and stationary plants and machinery is carried on regular basis at shorter intervals to minimize the amounts of hydrocarbons contributed to atmosphere. For construction of underground structures, ventilation systems are installed to ensure adequate supply of oxygen inside and to throw out noxious fumes and dust generated due to blast.

Wherever possible, fish passage is being provided especially in the projects located on the main rivers, so that migratory fish can access normal spawning areas upstream of diversion structure of a hydropower project. Recently, fish ladder has been provided in Larji Hydroelectric Project (126 MW), nearing completion on the river Beas in Himachal Pradesh. In projects having reservoirs, adequate facilities are being developed. Developers also pay reasonable sum depending on the capacity of the Project to the Department of Fisheries, Government of Himachal Pradesh for the development of fish hatcheries in the project area or in its vicinity.

Environmental flow, i.e. release of water to the river downstream from the diversion structure is mandatory. It is now being ensured that river reach between the dam and the power station has sufficient water flows so that fish and other aquatic organism are able to reside in the affected reach. Simple studies for maintenance of ecosystem downstream of the diversion structure are being conducted. These studies besides, consider amount of water required for villages located in the bypass reaches for irrigation, domestic and other uses.

Most of the projects being executed are run-of-the-river schemes with simple diversion structures. There are no major storage reservoirs in many of the schemes which are under execution or have been planned for future. At places where there is inundation, mostly confined within the river banks, vegetation cover is cleared before reservoir filling, in order to prevent generation of green house gases and disease vectors. There are no big reservoirs. Almost all the rivers and streams flow in step gradient with boulders in between or towards the banks, thus, there is no lack of dissolved oxygen in the river water downstream of the power house.

Social impact including loss of homes and agricultural areas are compensated as per resettlement and rehabilitation policy of the Government of Himachal Pradesh. It is being ensured that while executing hydroelectric projects in the state, people affected by the development of these projects are better off than before the construction of the Project.

Loss of forests and habitats for wildlife are compensated mainly by carrying out compensatory afforestation in the catchments area as per Catchments Area Treatment Plan for the river basin. Mitigation measures are incorporated in the design of power plants.

4.5 HYDROPOWER RESOURCES AND BENEFITS TO HIMACHAL PRADESH

Himachal Pradesh has hydroelectric power potential as one of its main resources. Initially most of the projects in the state were executed in the central sector. State was given a token share of the energy generated from these projects. Now, a new Hydel policy has been framed by the government of Himachal Pradesh. According to the policy, projects allotted to the private sector, at present 12 per cent free power is given to the state for the first twelve years and 18 per cent for the next eight years. After 40 years project is to be transferred to the state free of cost. Proper implementation of such policy would definitely benefit the economy of Himachal Pradesh not only in short-run but also in long-run.

Himachal Pradesh having hydropower as its most viable resource, can get benefited by issuing licence to the developers under private sector for thirty years duration and thereafter handling over of the project to the state free of cost, as per the international practices, so that hydropower resources of the state are retained by the state for the benefit of its future generations.

NOTES AND REFERENCES

1. The hydroelectric life cycle produces very small amounts of greenhouse gases (GHG). In emitting less GHG than power plants driven by gas, coal or oil, hydroelectricity can help retard global warming. Hydroelectric power plants don't release pollutants into the air. Hydroelectric enterprises that are developed and operated in a manner that is economical viable, environmentally sensible and socially responsible represent the best concept of sustainable development. That means, "development that today addresses people's needs without compromising the capacity of future generations for addressing their own needs" (World Commission on the Environment and Development,

1987). For further detail please see <water.usgs.gov/edu/hydroadvantages.html>.

2. For detail on this, please see, Zinta, R.L. and Tiwari, A.K. (2008), *Psychological Analysis on Project Affected and Likely to be Affected Families of Satluj Jal Vidyut Nigam Limited in Himachal Pradesh,* Institute of Integrated Himalayan Studies (UGC Centre of Excellence), Himachal Pradesh University, Shimla.

5

Hydroelectric Power Generation and Socio-economic Development

Since the present work is related to the study of hydroelectricity and economic development, it seems appropriate to present an overview of the socio-economic development of this Western Himalayan State of our country, i.e. Himachal Pradesh, which has immense hydroelectric potential. After considering the importance of development of power sector for the improvement of standard of living of the inhabitants and socio-economic development, sincere efforts have been made by planners, policy-makers in the state of Himachal Pradesh for rapid development of hydropower especially after the Sixth Five Year Plan (1980-85).

5.1 THE SOCIO-ECONOMIC PROFILE OF HIMACHAL PRADESH

Himachal Pradesh is the third largest Hill State of India, which is situated in the Western Himalayan region. This State came into being as a Part 'C' State of the Indian Union on 15th April 1948 after amalgamation of the 30 erstwhile Punjab and Shimla Hill States.[1] Then it comprised of four districts, namely Chamba, Mahasu, Mandi and Sirmaur. In 1954, the State of Bilaspur was merged with Himachal Pradesh and the number of districts rose to five. In 1960, one more district of Kinnaur was carved out of Mahasu district. On November 1, 1966, as a result of states' reorganization, certain parts of Punjab were transferred to Himachal Pradesh under the Punjab Reorganization Act, 1966, thereby adding three more districts, namely, Kangra, Kullu and Lahaul-Spiti. Thus its area nearly doubled. On 25 January 1971, it became a full-fledged state[2] of the Indian Union. However, two more districts—Una and Hamirpur were carved out of Kangra district and Mahasu district was divided into Shimla and Solan districts on September 1, 1972. Since then, there have been twelve districts in Himachal Pradesh viz. Bilaspur, Chamba, Hamirpur, Kangra, Kinnaur, Kullu, Lahaul-Spiti, Mandi, Shimla, Sirmaur, Solan and Una.

The geographical area of the state is about 55,673 square kilometers, which constitutes about 10.48 per cent of the total area of Indian Himalayan region and 1.69 per cent of the total area of India. It lies between 30° 22′ 44″ North latitude and 33° 12′ 40″ North latitude and 75° 45′ 55″ to 79° 04′ 20″ East longitude in the extreme north of India. It is surrounded by Jammu and Kashmir in the North and Uttaranchal in the Southeast. In the south and on the western side, the territory is bordered by Haryana and Punjab respectively, whereas on the east it borders with Tibet.

According to 2001 Census, total population of Himachal Pradesh stood at 60,77,900, which constitutes nearly 0.6 per cent of the total population of India. The decennial growth rate of population in the state during 1991-2001 was 17.54 per cent — less than that of all-India average (21.34 per cent). The maximum decennial growth rate of population was recorded in Solan district (30.64) and minimum decennial growth rate was in Lahaul-Spiti district (6.17). The highest rate of growth of population in Solan is probably due to extensive growth of industries and consequent in-migration of industrial workers. Kangra district stands first with population of 13,38,536 and accounts for 22.02 per cent of the total population of the state. About half of the population is concentrated in three districts, i.e. Kangra, Mandi and Shimla. Rural population constituted 90.20 per cent of the total population, considerably above the all-India percentage of 66.6. Density of population was 109 as against the all-India figure of 324. Whereas district Hamirpur accounted for the highest population density of 369, Lahaul-Spiti had the lowest density of 2 only.

The sex ratio, which is the number of females per thousand males, was 968 and it was higher than all-India ratio of 933. The trends in sex ratio[3], which, however, emerge from the inter-censal data, reflect in a substantial way the changes that could be occurring in attitudes towards the girl child and the general biases towards the females and their empowerment. The highest sex ratio was recorded in Hamirpur district at 1099 and the lowest 802 in Lahaul-Spiti.

The total workforce of the State increased by 66 per cent during the decades 1981-2001. The percentage of males in the total workforce decreased from 62.93 per cent to 56.33 per cent, whereas, percentage of females in the total workforce registered an increase from 37.07 per cent to 43.66 per cent during the same period. The occupational distribution of the population shows that, as per 2001 census, the percentage of the main workers to total population was 49.22 and it was marginally higher than the corresponding all-India percentage of 39.19. Out of them, 65.33 per cent were classified as cultivators, 3.15 per cent as agricultural labourers, 1.75 per cent were engaged in household industries, and 29.77 per cent were other workers. Himachal Pradesh is predominantly an agricultural economy with about 69 per cent of its total main workers engaged in this sector. Agriculture is not only a profession of the people but also a way of life supporting directly and indirectly about 90.20 per cent of the rural population.

The state is characterized by are a wide variety of climatic conditions, topography and social features. It can be divided into three climatic zones, namely, the Outer Himalaya or the Shiwalik, the Inner Himalayas or Mid-mountains, and the Alpine zone or The Greater Himalaya.

The first zone consists of the lower hills of districts of Kangra, Una, Hamirpur, Bilaspur, Sirmaur, Solan and Mandi.

The Shiwalik in Himachal Pradesh is also known by several local names like Dhog Dhar in Sirmaur, Ram Garab Dhar in Una, Chaumuke range, Dharti Dhar, Sikander Dhar and Naina Devi Dhar in Bilaspur district. This range is composed of highly unconsolidated soils, which has made it prone to erosion. The annual rainfall in this zone varies between 150 cm. and 175 cm. The altitude of this zone ranges from 350 metres to 1500 metres above mean sea level whereas the soil is shallow and loam to clay and as such is suitable for agricultural production.

The second zone, which has altitude from 1500 metres to 4500 metres above mean sea level, consists of upper areas of districts of Sirmaur, Mandi, Kangra, Shimla and Chamba. More particularly, the areas that fall in this zone are upper areas of the tehsils of Pachhad and Renuka in Sirmaur district; Chachiot and Karsog tehsils of Mandi district; Palampur tehsil and upper parts of Kangra district; upper Shimla hills, and upper parts of Churah tehsil of Chamba district. The rainfall varies between 75 cm. and 100 cm. and the variety of soil ranges from silty loam to clay loam of dark brown colour. These types of soil characteristics, climatic conditions and topography are conducive for the cultivation of commercial crops, off-season vegetables and horticultural produce, etc.

The Area falling at an altitude of more than 4500 metres above mean sea level is termed as the Alpine zone. This Great Himalayan range runs along the Eastern boundary and is cut across by the Satluj river. The range separates the drainage of the Spiti river from that of the Beas. This zone comprises Kinnaur district, Pangi tehsil of Chamba district and some areas of Lahaul-Spiti district. This zone remains snowbound for about six months in a year. Here, the rainfall is very scanty and soil is of high texture with variable fertility. The climate is temperate in summer and semi-arctic in winter. This kind of climate and soils are best suited for the cultivation of dry fruits.

Total area under forest is about 10,93,545 hectare[4], which was about 24.0 per cent of the total reporting area in 2000-01, whereas the corresponding all-India percentage was 21.9 per cent. Kullu district has the highest forest cover in the state, i.e. about 61.35 per cent of its geographical area, in contrast to district Kinnaur which has only 6.58 per cent forest cover. According to Himachal Pradesh Forest Report, 2001, per capita forest area, which was reported to be 0.93 hectare in 1966-67, has decreased marginally to 0.74 hectares. Depletion of forest cover has become cause of concern. Satellite imageries show that percent of forest cover is much lower than shown in government's record. Efforts are being made to bring maximum area under green cover by implementing State's own afforestation projects, Government of India's projects and also through external aided projects. The World Bank has sanctioned a sum of Rs. 540 crores for Integrated Watershed Development Project for the Mid Himalayas (India 2006, p. 964).

During the last two decades, the economy of Himachal Pradesh has witnessed a steady structural change. From the perusal of Table 5.1, it is evident that the share of agriculture including forestry and fishing in the Net State Domestic Product (NSDP) of Himachal Pradesh gradually declined from 50.35 per cent in 1980-81 to 25.5 percent in 2000-01. During this period, there took place an ample increase in the share of secondary sector from 18.70 per cent in 1980-81 to 33.23 percent in 2003-04. The maximum growth has taken place in the secondary sector. This structural change from Primary to Secondary and Tertiary sectors is a healthy sign of a growing economy.

TABLE 5.1
Sector-wise Net State Domestic Product at Current Prices in Himachal Pradesh

Sl. No.	*Sector*	*1980-81*	*1990-91*	*2000-01*	*2003-04*
1.	Primary	50.35	38.31	25.50	26.38
2.	Secondary	18.70	24.07	32.00	33.23
3.	Trade and Transport	7.70	8.30	11.20	14.28
4.	Finance and Real Estate	8.24	8.67	9.10	7.56
5.	Community and Other Services	15.02	20.64	22.30	18.55
	Total	100.00	100.00	100.00	100.00

Source : State Domestic Product of Himachal Pradesh, 1980-81 to 1992-93 and 1993-94 to 2000-01, p. 38 and 17 respectively and Economic Survey, Himachal Pradesh, 2004-05, pp. 7-9.

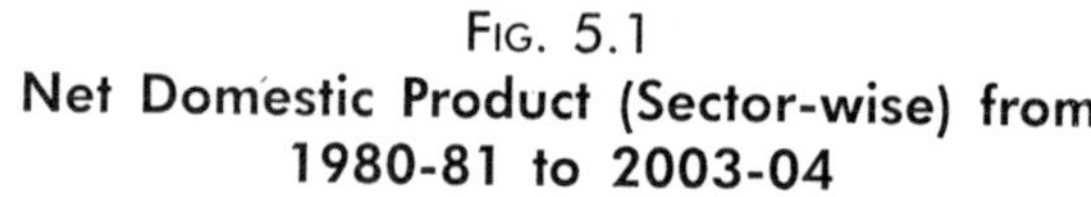

FIG. 5.1
Net Domestic Product (Sector-wise) from 1980-81 to 2003-04

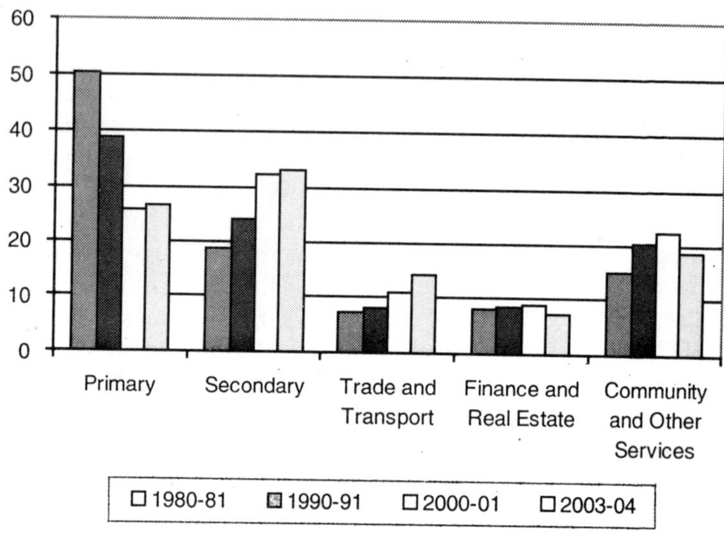

It may be observed that significant changes appear to have taken place in the economic structure of the state during 1980-81 to 2003-04. Despite a declining share of primary sector in the state's domestic product agriculture sector still continues to be the single most important sector whether we see its contribution to the State's income or its labour absorbing capacity.

The developing economy of Himachal Pradesh has experienced uneven spatial pattern of development in the post-independence period. The district-wise per capita income of Himachal Pradesh for the year 2000-01 indicates that district Solan enjoyed the first position among all the districts followed by Lahaul-Spiti and Kinnaur. On the other hand, lowest per capita income was recorded by Hamirpur followed by Una.

Agriculture in the state exhibits the blend of both progressive and traditional farming. The state has made spectacular headway in the production of apples and other temperate fruits in the higher hills, and vegetable crops and vegetable seeds as well as sub-tropical fruit in lower hills.

TABLE 5.2

District-wise Domestic Product and Per Capita Income in Himachal Pradesh (2000-2001)

Sl. No.	*Districts*	*Net Domestic Product** *(Rs. in lakh)*	*Per Capita Income* *(Rs.)*	
1.	Bilaspur	77133	22229	(5)
2.	Chamba	81490	17644	(8)
3.	Hamirpur	52153	12020	(12)
4.	Kangra	225917	16370	(9)
5.	Kinnaur	23858	28449	(3)
6.	Kullu	75206	21155	(7)
7.	Lahaul-Spiti	13115	35379	(2)
8.	Mandi	140244	15375	(10)
9.	Shimla	199720	27548	(4)
10.	Sirmaur	96832	21709	(6)
11.	Solan	160289	35692	(1)
12.	Una	56309	12658	(11)
Himachal Pradesh		1331996	19784	

*2001-02.

Note : Figures in parentheses indicate ranking of the district.

Source : Economics and Statistics Department, H.P. (Unpublished data obtained from official files).

FIG. 5.2

District-wise Per Capita Income

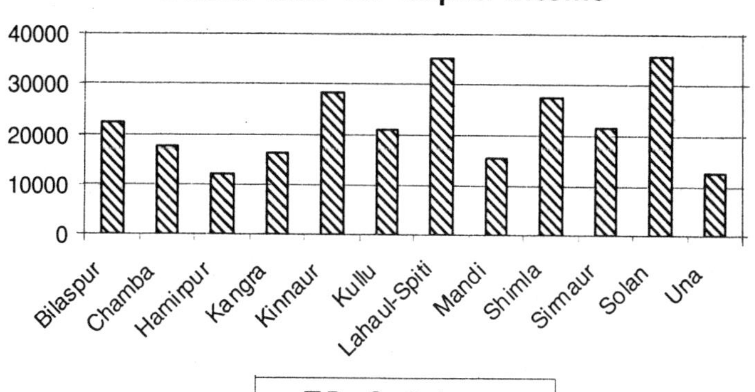

Agriculture and allied activities (including forestry and logging) continue to be the mainstay of the population as these provide livelihood to about 71 per cent of the working population. Income from agriculture and allied sectors accounts for nearly 21.7 per cent of the total State Domestic product. The soils of this state are not as fertile as that of the neighbouring states of Punjab and Haryana. It is more suitable for food crops like maize, rice, barley and wheat, and for fruit-cultivation in cold regions. Only about 17.2 per cent of the net sown area is irrigated which is mainly in the form of snow fed gravity flow channels (kuhls). About 88 per cent of the gross cropped area is under food grains. The average holding size comes to 1.2 hectares while 84.5 per cent of the total holdings are small and marginal. The state receives an average rainfall of 152 cm. and about 80 per cent of the area is rain-fed. The food-grains production in 2004-05 was 16.36 lakh tonnes.

The net sown area of this state was 554 thousand hectares and it constituted about 12.2 per cent of the total reporting area in 2000-01. This ratio is much lower than the all-India figure of 43.4 per cent. The economy of the State is Agriculture-based and majority of the population has been sustaining on the agricultural output as 96.7 per cent of the total area sown under all crops is covered by food crops and only 3.3 per cent by non-food crops. The food crops constitute 99.5 per cent and non-food about 0.5 percent of the total out-turn of the crops grown. Under the food crops, it is the cereal crops, viz. wheat, barley, maize, paddy and other cereals which account for major share of area as well as production of the crops sown. The cereal crops contribute 82.7 per cent of the total cropped area and 81.0 per cent to the total output. Keeping in view the peculiar physical and agro-climatic conditions of the state, the scope to bring more area under cultivation is severely limited and available land is exhaustively used for raising cereal crops for the purpose of self-sufficiency of the rural masses.

The topography of this hill state renders it difficult to use modern agricultural implements in high hill and mid-hill regions. In low and valley areas, however, mechanization of agriculture has steadily increased with greater use of iron ploughs, electric pumps, tractors, sprayers, etc. In some parts of the state farmers are increasingly adopting the improved 'seed-irrigation-fertilizer-pesticide technology' and the system of multiple cropping. Consequently, crop productivity has increased in these areas.

Nature has endowed Himachal Pradesh with a wide range of agro-climatic conditions, which have helped the farmers to cultivate a large variety of fruits ranging from temperate to sub-tropical. The main fruits under cultivation are apple, pear, peach, plum, apricot, citrus fruits, mango, litchi, guava and strawberry, etc. The total area under fruit cultivation, which was only 792 hectare in 1950 increased to 2.23 lakh hectares in 2004. Similarly, the fruit production has also increased from a bare 1200 MT in 1950 to as much as 6.85 lakh tonnes in December 2004. Horticulture[5] generates a huge gross domestic income of about 1000 crore annually.

Industrial development has been given a big boost in the State. Pollution free environment, abundant availability of power and rapidly developing infrastructure, peaceful atmosphere, responsive and transparent administration are some of the added attractions and advantages that the entrepreneurs get in Himachal Pradesh. 229 large and medium, and about 31,384 small scale industrial units with an investment of about Rs. 3,280.99 crore have been set-up in the state. The sector is contributing 14 per cent to the State Domestic Product and the annual turnover on this account is about Rs. 6000 crore (India 2006, p. 962). As stated earlier, the relative contribution of secondary sector to the state's NSDP has increased substantially since 1980-81 due to concerted efforts of state and central governments towards the industrial development. Nevertheless, the pace of industrial development is rather slow due to limited size of the market, locational disadvantages in terms of long distances from markets, prohibitive costs, etc. Notwithstanding these constraints, industrial areas at Parwanoo, Barotiwala, Baddi, Poanta Sahib, Mahatpur, Shamshi, Nagrota Bagwan, Bilaspur, Recong Peo, Sansarpur Terrace; electronic complexes at Solan, Mandi, Kala Amb, Hamirpur, Shoghi and Chamba; and industrial estates at Solan, Dharampur, Kangra, Jawali and Dehragopipur have been established in order to accelerate the pace of industrialization and to tackle the problem of growing unemployment. The number of factories per lakh of population rose from 24 in 1980-81 to 33 in 2000-01. Similarly, the number of factory workers per lakh of population grew from 314 in 1980-81 to 1247 in 2000-01. The corresponding all-India figure was 892 in 2000-01. The lower number of factory workers per lakh of population may be attributed to the fact that most of the industrial units located in this state belong to the category of small scale industries employing limited number of workers.

Roads are the lifeline and major means of communication in the predominantly hill State of Himachal Pradesh. Out of its 55,673 sq. km. area, 36,700 sq. km. is inhabited and its 16,807 inhabited villages are scattered over slopes of numerous hill ranges and valleys. Due to a high priority accorded to construction of roads during the first four plans, the contribution of trade and transport, though low, has shown a steadily rising trend. When the state came into existence in 1948 there were just 288 km. of roads, which figure has gone up to 28,588 km. in 2004. During 1980-81 the road density in Himachal Pradesh was just 22.64 kilometers per 100 square kilometers and it increased to 47.4 km. in 2000-01, which continues to be less than the all-India average of 76.8 kms. When compared with that of other Himalayan hill States, it is higher than that in Jammu and Kashmir (5.9 kms.), Meghalaya (37.4 kms.), Mizoram (32.8 kms.), Sikkim (25.8 kms.), Arunachal Pradesh (12.2 kms.) but lower than that in Nagaland (82.8 kms.) and Tripura (140.4 kms.).

Present level of communication facilities in the state is now well comparable to all-India average. Due to favourable policy of the central government towards provision of means of communication in hilly areas, the existence of post offices per lakh of population in 2001 was 46 while it was only 19 at all-India level. There is enormous increase in tele-density. The number of telephones/cellular phones per lakh of population in Himachal Pradesh was 8120 in 2001 while at all-India level it was 4890. Under the New Telecom Policy, 1999, with provision of affordable and effective communication as its core vision and goal, the telecommunication sector in India has achieved a lot in recent years. The total number of telephones (basic and mobile) rose from 22.8 million in 1999 to more than 125 million at the end of December 2005 (*Economic Survey* 2005-06, p. 182).

Himachal Pradesh has also achieved remarkable growth in tele-density after implementation of the New Telecom Policy, 1999.

The banking facilities in terms of population and area coverage have also been rapidly increasing. The number of bank offices per lakh of population was just 1.3 in June 1969 at the time of nationalization and it increased to 12.94 in December 2001. The corresponding all-India figures were 1.6 and 6.57 respectively. However, the credit-deposit ratio was only 21.7 per cent in December 2001 in Himachal Pradesh and it considerably lagged behind the all-India ratio of 56.71 per cent. Thus, the growth of banking in terms of numerical/geographical expansion has been far more than that in terms of credit deployment.

Cooperation has aptly been described as a movement rather than a mere programme. Its objectives cover acceleration of economic growth coupled with social justice. It is conceived as an important factor in building up an egalitarian and non-exploitative economic and social order. Cooperative movement in Himachal Pradesh presents a picture of adequate coverage both village-wise and family-wise. The number of primary agricultural credit societies (PACs) in 1980-81 was just 2166 which rose to 4332 in 2004-05. Loans disbursed by the societies per cultivator/member were Rs. 131.00 in 1980-81, which increased by about twenty-four fold to Rs. 3178.60 in 2004-05.

The per capita consumption of electricity, which was merely 74.78 kwh in 1980-81, rose to 279 kwh in 2000-01, yet it was significantly less than the all-India average consumption of 338 kwh. Also, the number of villages electrified increased from 24 per cent in 1970-71 to 100 percent in 1990-91. This percentage is much above the all-India percentage of 86.32 in 2000-01. Five perennial rivers—Yamuna, Satluj, Beas, Ravi and Chandrabhaga or Chenab flow through this state. Although these are not of much importance so far as irrigation is concerned, yet a huge hydroeclectic potential exists which is estimated at around 25,000 MW. Besides this, these rivers act as a crucial source of drinking water. Thus far, out of the total hydel potential, only 3363 MW has been harnessed, out of which Himachal Pradesh accounts for just 272 MW. Out of total hydroelectric potential of 97,000 MW in the country, about 32 per cent lies in north-western Himalayan region. If this potential is exploited fully, the entire demand for Northern India would be met and it will give a boost to industrialization. However, as the Government has embarked upon an accelerated Power Development Programme; the State is speedily moving towards a 'Power State' of the Country.

The social sector development in the State has been progressive in absolute terms throughout the planning period. Himachal Pradesh has a good track record of development in the field of education, particularly in elementary education and education of girls. The development experience of various nations overtime has led to the conviction that literacy and education have a direct role in human development and are instrumental in facilitating other achievements including economic prosperity. The State has about 15,000 educational institutions, including three universities, two Medical Colleges, one Engineering College in the Government Sector and a number of technical, professional and other educational institutions. According to 2001 census literacy rate in the state was 77.13 per cent and the female literacy rate was 68.08 per cent, whereas corresponding all-India average of literacy rate was 65.38 per cent. The female literacy in this state, which was 20.23 per cent in 1971, has risen to 68.08 per cent in 2001 showing more than three-fold increases in the literacy of females. This has become possible with the help of well-directed public intervention in support of social opportunities, active agency of women, and local democracy and social cooperation. Himachal Pradesh has benefited from a strong state commitment to development, particularly focused on the social infrastructure in rural areas. Roads and schools, in particular, have been among the government's highest priorities. In spite of the adverse topography and settlement pattern (with small hamlets scattered over wide areas in a difficult terrain), enormous progress has been made in the provision of basic amenities at the village level (Jean Dreze and Amartya Sen, pp. 104-05). In literacy, Himachal Pradesh occupied tenth place in the ranking among the Indian states. The dropout rates (I-VIII) for Himachal Pradesh stand at 28.5 per cent for boys and 28.1 per cent for girls against the all India figures of 54.4 per cent for boys and 60.1 per cent for girls.

Hamirpur district ranked the highest with a literacy rate of 83.16 per cent, followed by Una (81.09 per cent), and Kangra (80.09 per cent), but this ratio was the lowest in Chamba district, (63.73 per cent) followed by Sirmaur, Lahaul-Spiti, Kullu and Mandi. Male and female literacy rates of the state were 87.13 per cent and 70.53 per cent respectively in 2001. According to the National Family Health Survey 1998-99, school participation rates among 6-14 years old in Himachal Pradesh are as high as 99 per cent for boys and 97 per cent for girls. In this respect Himachal Pradesh is at par with Kerala and well ahead of all other States.

The state also appears to have performed better in terms of provision of medical facilities. The number of medical institutions per lakh of population increased steadily from 17.24 in 1980-81 to 29.24 in 2000-01, and this figure was considerably higher than the all-India figure of 5. Similarly, the number of beds per lakh of population rose from 59 in 1980-81 to 173 in 2000-01, as against the all-India average of 7. However, the provision of various facilities also needs to be viewed in terms of area keeping in view the low density of population, especially in the tribal belt. Nevertheless, growth of medical facilities is reflected in reduction of crude birth and death rates and significant improvement in the infant mortality rate. The birth and death rates per thousand in 2001 were 21 and 7 respectively in Himachal Pradesh whereas all-India averages were 25.8 and 8.5 respectively.

Provision of safe drinking water is one of the important indicators of social development. According to 1981 census, there were 16,807 inhabited villages in Himachal Pradesh out of which 11,887 villages were not provided with drinking water facilities. By 31st March 1995, drinking water facilities had come to be provided to all the census villages. During 1991-93 the Government of India had conducted a status survey in respect of Rural Water Supply. According to this survey, although census villages are main habitations but all the small habitations are not connected with drinking water schemes. The total number of main habitations and other habitations are 45,367 out of which still 2738 small habitations had not been covered under safe drinking water schemes as on 1.4.2000.

However, the state level indicators do throw light upon the process of socio-economic development and standard of living which people of this enjoy partly due to rapid exploitation of hydroelectric power in the state of Himachal Pradesh as compared to other similar placed hill states as well as the national average.

Notes and References

1. These 30 small princely states were Bhagat, Bhajji, Baghal, Beja, Balsan, Bushahr, Chamba, Darkoti, Delath, Dhadi, Dhami, Ghund, Jubbal, Khaneti, Koti, Kumarsain, Kunihar, Kuthar, Madhan, Mahlog, Mandi, Mangal, Ratesh, Keonthal, Rawingarh, Sangri, Sirmaur, Suket, Tharoch and Theog.
2. Himachal Pradesh maintained its status of Part 'C' state of the Indian Union till conferment of full-fledged state on 25th January 1971. State Reorganisation Commission(1956) headed by Mr. justice Fazil Ali and two other members Mr. H.N. Kunzru and Mr. K.M. Panik, recommended the merger of all part 'C' states either to adjoining states or to maintain these independently as union territories for a specified period. The chairman of the Commission was not in favour of making Himachal Pradesh a separate state on the ground that due to prevalence of gross backwardness in the state, the backwardness will further increase if full-fledged statehood was conferred. Nevertheless, full-fledged statehood was granted and presumption of the Chairman was proved incorrect as the state of Himachal Pradesh has achieved notable levels of economic and social development and now well comparable with many developed states.
3. Himachal Pradesh is one of the few States in the country where the gender equity is an integral part of the social ethos as well as the stance on development programmes. This is amply evidenced by the fact that the State has a high sex ratio of 970 females per thousand males. The female literacy is way above the national level. The State also has a high incidence of women employment than most States of the country (Annual Plan, 2003-04, pp. 39-40).
4. The total geographical area of the state is 55,673 sq. km. As per record, the total forest area is 37,033 sq. km. Out of this, 16,376 sq. km. area is not fit for tree growth comprising alpine pastures, area under permanent snow, etc. The cultivable recorded forest is only 20,657 sq. km.
5. During the Tenth Five-Year Plan, Horticulture Technology Mission for the integrated development of Horticulture has been implemented. This mission is based on the "end of end approach" taking into account the entire gamut of Horticulture Development with all backward and forward linkages in holistic manner. Under this scheme four centres of excellence are being created in different Agro Climatic Zones with common facilities like water harvesting, vermicompost, greenhouses, organic farming and farm mechanization.

6

Summary and Conclusions

The main thesis of the study was to explore the advantages and disadvantages of hydropower projects on environment and quality of life. As an introduction to the study (in Chapter 1), hydroelectricity scenarios have elaborately been discussed both for world and India levels. It is unanimously accepted that electricity is most helpful component of modern civilization and is sometime called as man's most useful servant. Hydroelectricity is undeniably the largest renewal resources used for electricity generation. It uses the energy of running water, without reducing its quantity, to produce electricity. Therefore, all hydroelectric developments, of small or large size, whether run of the river or of accumulated storage, fit the concept of renewable energy.

Hydroelectric power plants with accumulation reservoirs offer incomparable operational flexibility, since they can immediately respond to fluctuations in the demand for electricity. The flexibility and storage capacity of hydroelectric power plants make them more efficient and economical in supporting the use of intermittent sources of renewable energy, such as solar energy or Aeolian energy.

River water is a domestic resource which, contrary to fuel or natural gas, is not subject to market fluctuations in addition to this, is the only large renewable source of electricity and its cost-benefit ratio, efficiency, flexibility and reliability assist in optimizing the use of thermal power plants. Hydroelectric power plant reservoirs collect rainwater, which can then be used for consumption or for irrigation. In storing water, they protect the water tables against depletion and reduce our vulnerability to floods and droughts. The operation of electricity systems depends on rapid and flexible generation sources to meet peak demands, maintain the system voltage levels, and quickly-re-establish supply after a blackout. Energy generated by hydroelectric installations can be injected into the electricity system faster than that of any other energy source. The capacity of hydroelectric systems to reach maximum production from zero in a rapid and foreseeable manner makes them exceptionally appropriate for addressing alterations in the consumption and providing ancillary services to the electricity system, thus maintaining the balance between electricity supply and demand. The hydroelectric life cycle produces very small amounts of greenhouse gases. Hydroelectric power plants don't release pollutants into air. They very frequently substitute the generation from fossil fuels, thus reducing acid rain and smog. In addition to this, hydroelectric developments don't generate toxic by-products.

The worldwide theoretical, technical and economic potential of hydropower have been estimated as theoretical potential at about 40,500 Tera Watt Hours, technical potential at about 14,320 TWh, and economic potential at about 8100 TWh. Currently, about one-third of the economic potential has so far been developed. The US and Western Europe have developed about 76 per cent and 65 per cent of their potential respectively. Whereas countries like Africa and Asia have only developed about 20 per cent of their potential. The world installed hydro capacity stands at 6,94,000 MW. Canada, Brazil, US, China and Russia were the top five producers of hydroelectric power and their collective hydropower generation accounted for 49 per cent of the world total. Worldwide about 125,000 MW of hydro-capacity is under construction, bulk of which is in the developing countries. China, India, Malaysia and other developing Asian countries have striving plans of hydropower capacity additions.

India has been classified into six major river systems, namely, Indus, Brahmputra, Ganga, Central Indian River System, East flowing River System for the purpose of hydro-electric potential survey. These river systems have been further divided into 49 basins. The first systematic and comprehensive study to assess the hydro-electric resources in the country was undertaken during the period 1953-59 by the Power Wing of the erstwhile Central Water and Power Commission. These studies placed the economical exploitable hydropower prospective of the country at 42,100 MW at 60% load factor, similar to an annual energy generation of 221 billion units.

The re-assessment studies, accomplished by Central Electricity Authority in 1987, have placed the hydropower potential at 84,044 MW at 60% load factor. A total of 845 hydro-electric schemes have been identified in the various basins, which are annually projected to yield about 442 billion units of electricity. A vision paper highlighting comprehensive approach for development of 1,50,000 MW of Hydro Power matching to the assessed potential of 84,044 MW was prepared by CEA and submitted to Ministry of Power during March 2001.

The installed power generation capacity in India has increased significantly from 1400 MW in 1947 to 1, 18,419.09 MW as on 31 March, 2005 comprising 80,902.45 MW thermal, 30,935.63 MW hydro, 38, 11.01 MW wind and 2770 MW nuclear. A capacity addition programme of 6344.52 MW was fixed for the year 2005-06. Considering the fact that a large chunk of proportion of the installed capacity will come from the public sector, the outlay for the power sector was raised from Rs. 45,591 crore during the Ninth Plan to Rs. 1,43,399 crore in the Tenth Plan. This included a gross budgetary support of Rs. 25,000 crore and the remaining Rs. 1,18,399 crore was from internal and extra budgetary resources. Power generation during 2004-05 was 587.366 BUs comprising 486.031 BUs thermal, 84.947 BUs hydro. The target of Power generation for 2005-06 was fixed at 621.500 BUs.

Hydroelectricity and quality of life was the subject matter of Chapter 2. It is observed that Electricity is a vital infrastructure for economic development. There is a close correlation between the per capita GNP of a country and the per capita electricity it consumes. It is a well established fact that electricity and economic growth go hand in hand. In fact, some economists regard electricity as the fourth factor of production, in addition to the traditional listing of land, labour and capital. Some of the major areas where electricity has substantial impact on development are mining, irrigation, transportation, production of consumer goods, cement, steel, domestic appliances, recycling of waste materials into useful ones, etc.

Accelerating economic growth and achieving higher standards of living (quality of life) depend upon the availability of adequate and reliable power at an affordable price. Unlike other commodities, electricity cannot be stored for future use. In other words, its generation and consumption have to be simultaneous and instantaneous. Furthermore, electricity, a major source of energy, is regarded as an important factor for bringing about radical change in the socio-economic life of a community. Because of multifarious uses of electricity, such as for lightning and as a source of motive power, its introduction does not merely facilitate provision of better amenities but augments productive capacity in different sectors of the economy through its wide range of applications. The availability of electric power affects industries in two principal ways. Firstly, the traditionally operated industries gradually change over to the use of electric power, and secondly, new industrial units operated by power come up. Use of electric power, therefore, strengthens and expands both small and large industrial sectors and makes them more efficient and productive.

Rural electrification has a significant and positive impact on agricultural production, thereby creating additional processible resources. Rural Electrification Programme is not to be seen as a drive to light the house of the people. Two related aspects should be emphasized in this context. One is the problem of finding alternative sources of energy as hitherto complete dependence on forests cannot continue for long for obvious reasons. Therefore, the rural electrification scheme has to be viewed as a part of the programme for a search of alternative sources of energy and its appropriate utilization. The second aspect is related to the direct benefits of electricity in underdeveloped rural areas.

Therefore, power development has a significant role in promoting socio-economic development by creating a base on which a higher level of economic activity can be carried out. Energy is like blood in human body. Deficiency of blood in human body reduces not only physical power but also mental ability. In the similar way adequate energy supply is a *sine-qua-non* for rapid and sustainable economic development. Social and economic aspects of population, especially those who reside near the project site, are one of the most important considerations for any development project. The construction phase of a project generally has a pronounced impact on the socio-economic aspects on the local population for the reasons that a section of population faces displacement from their land which is acquired for the construction of project and the construction activities brings in a sea change in the local environmental settings due to large scale influx of population altering the traditional economic activities, etc. The construction sites of hydropower projects have shown inward migration of large number of labour force including drastic change in the socio-economic fabrics of the local population

The state of Himachal Pradesh has immense hydro-potential in its five river basins, Chenab, Ravi, Beas, Satluj and Yamuna which emanate from the western Himalayas and pass through the State. Himachal Pradesh has hydel potential of over 80,000 MW which comes to about one-fourth of that of the entire country, but only 21,346 MW have been identified so far. Out of this potential only 6060 MW stands harnessed so far. Issues in speedy development of hydropower Projects have also been briefly traced theoretically in this chapter. Under the section of development of hydropower potential under Five Year Plans, it was observed that power sector has remained as a priority sector. Consequently, the installed capacity which was pitiable at 2.000 M.W. in 1950-51 had increased to 48.919 M.W. in 1970-71, and further to 126.520 in 1980-81 and 326.300 MW in 2002-03. The installed capacity of power sector has expanded by 163 times from the installed capacity that existed during 1951. In the matter of electrification of inhabited villages, Himachal Pradesh has made remarkable advancement. In 1970-

71 only 24 per cent of inhabited villages were electrified. By 1980-81 this percentage went up to 54.9. By 2004-05, it has achieved 100 per cent electricity availability for its households.

The analysis of generation of electricity has special significance in the present context because Himachal Pradesh happens to be one of the richest areas in the country in the matter of hydro-electric potential. In 1970-71 the generation of electricity was negligible. The power generation which was 2450.66 MW in 1980-81 touched the level of 12623.79 in 1990-91. The total electricity generation in the year 2003-04 touched the level of 13569.50 MW. During the thirty three year period between 1970-71 to 2003-04, while electricity generated increased by as much as 26 times, electricity consumption also increased by more than 24 times.

The assessment of extent of relationship between hydro-electricity and economic development in Himachal Pradesh has shown positive and significant association between pairs of indicators, viz. Net State Domestic Product and hydropower generation; Net State Domestic Product and consumption of power; Per capita income and hydropower generation; Per capital income and consumption of power; Hydropower generation and industrialization; Hydropower consumption and industrialization, etc.

The River Satluj, which is one of the key basins featuring in the hydro development plan of the state of Himachal Pradesh, rises in the Tibetan Plateau (Rakastal-Mansarovar lake; at an elevation of about 4570 m above mean sea level), travels about 1450 km (320 km in China, 758 km in India, and 370 km in Pakistan) before it meets the Chenab River and subsequently the Indus. Governments of Himachal Pradesh and India are working to exploit the full hydro-potential of the Satluj river Basin though both private and public developers. Some of the projects proposed for construction on this river are Khab and the 1000 MW Karchham Wangtoo project upstream of Rampur and 425 MW Luhri and 800 MW Kol dam projects down stream. The 1500 MW Nathpa Jhakri HEP, immediate upstream is already in stage of operation. The most celebrated dam on the river is the Bhakra Dam, which was completed in 1963. Downstream of Bhakra too there are structures on the river, including the Nangal diversion dam and Ropar barrage.

The description of Hydro-Electric Power Projects on Satluj River Valley in Himachal Pradesh was given in Chapter 3. The catchments area of the Satluj at Rampur is about 50,800 km^2 (49,800 km^2 at Nathpa Dam), of which about 30% falls in India and the remainder in China. The rivers in the catchments are fed by snow melt, particularly in China. A small portion of the project catchments also receive precipitation due to the South-West monsoon (June-September). The peak flows of the river occur during June to September, while the lean period occurs between October and April. Much hydrological study of the Satluj has previously been performed in preparation for the construction of the existing Bhakra Dam, which is downstream of Rampur, and for the construction of the upstream existing Nathpa Jhakri scheme. Water availability studies were carried out from 1963 onward by using observed discharges at Rampur town and river diversion works will be designed to withstand a river flow corresponding to a 10,000 years return period flood, which has been assessed to be 7,151 cubic meters per second at Rampur.

The Satluj Jal Vidyut Nigam Limited—SJVN (formerly Nathpa Jhakri Power Corporation Limited—NJPC) was incorporated on Mary 24, 1988 as a joint venture of the Government of India (GOI) and the Government of Himachal Pradesh (GOHP) to plan, investigate, organize, execute, operate and maintain Hydroelectric power projects in the river Satluj basin in the state of Himachal Pradesh and at any other place. The present authorized share capital of SJVNL is Rs. 4500 crore. The Nathpa Jhakri Hydroelectric Project (1500 MW) was the first project undertaken by SJVN for execution.

SJVN, contemplates to become a 4000 MW company by the end of 11th Plan, for which a number of projects in the states of Himachal Pradesh and Uttarakhand, have been taken up for preparation of DPR and subsequent execution:

In the State of Himachal Pradesh

(1) Luhri Hydel Project (775 MW) on river Satluj, in district Shimla of Himachal Pradesh, India.
(2) Khab HE Project (1020 MW) on river Satluj, in district Kinnaur of Himachal Pradesh, India

In the State of Uttarakhand

(1) Devsari Dam Hydel Project (252 MW), on river Pindar, loated in district Chamoli, of Uttarakhand, India.
(2) Naitwar Mori HE Project (34.5 MW), on river Tons (a tributary of river Yamuna), located in district Uttarkashi of Uttarakhand, India.
(3) Jakhol Sankri HE Project (33 MW), on river Supin, located in district Uttarkashi, of Uttrakhand, India.

Nathpa-Jhakri Project on river Satluj with an installed capacity of 1500 MW (6 x 250 MW) is the biggest hydroelectric project of the country. The project comprises of a 62.5 m high concrete dam, an underground desilting complex for eliminating particle size of 0.2 mm and above comprising four champers, each 525 m long, 16.31 wide and 27.5 m deep; 27.30 km long head race tunnel of 10.15 m diameter designed for 405 m^3/sec and a tailrace tunnel of 10.15 m dia and 982 in length. The project has been executed by Satluj Jal Vidyut Nigam (Formerly Nathpa Jhakri Power Corporation), a joint venture of Government of India and Government of Himachal Pradesh, the two sharing cost of the project in the ratio of 3:1 respectively. The design energy from the project works out to 6951 MU in a 90% dependable year and 7351 MU in an average year. Power is being transmitted to Northern Grid through a double circuit of 400 kV transmission lines. Himachal Pradesh, Punjab, Haryana, Chandigarh, Delhi, Uttar Pradesh and Jammu & Kashmir are the beneficiary States. Himachal Pradesh gets 12 per cent of energy free of cost as royalty besides getting its share in energy of 25 per cent by virtue of equity participation. The first Unit of the Project started commercial generation from October 6, 2003 and the last Unit was commissioned on May 18, 2004. The World Bank provided a loan of 437 million US dollars through Government of India for generation component of this project which has been paid back. Part of the cost for supply of electrical and hydro-mechanical equipment was financed by raising loans from the Indian Financial Institutions besides bilateral funding from European markets.

A study was conducted on Impact Assessment of Resettlement and Rehabilitation Plan of Nathpa-Jhakri Hydroelectric Power Project of Himachal Pradesh. The study compared the magnitude of indices during 2002 (after the NJHEP project implementation) with the base line study (the situation as it existed in 1996) and with the control sample household data (households in project area which were not affected by NJHEP). The findings of the study were:

- Family size of the PAF's declined from 7.14 to 5.44 persons per family.
- Literacy rate has increased from 58% to 73%.
- Average annual household income increased from Rs. 21,648 (at 1996 price level) to Rs. 76,575 which works out to be Rs. 29,114 in 1996 and Rs. 10,4,640 at 2002 price levels.
- Proportion of families living below poverty line has decreased from 25.6% to 16.8%.
- Average per capita monthly expenditure of PAF's increased from Rs. 575 to Rs. 674, showing an improvement in their consumption pattern and standard of living.
- Percentage of workers engaged in regular employment increased from 20% to 30%.
- Percentage of workers engaged in business increased from 7% to 9%.
- More people live in *Pucca* houses (45% as compared to 21% earlier).
- More people have separate bathrooms in their houses (46% as compared to 21% earlier).
- Improvement in the overall living standards of the families due to NJHEP R&R Plan implementation.
- Diversification of income and employment avenues through income generation schemes, towards business and other self-employment activities is taking place.
- Improvement in the housing standards, quality of health care due to enhancement of diagnostic facilities with the introduction of mobile health unit by SJVN.

- SJVN has taken measures to strengthen the existing infrastructure facilities, including health facilities, education and roads which are providing immense benefits to the PAF's in the project area.
- Project affected families have received full and adequate compensation. The compensation amount has been used rationally and judiciously by PAFs. Overall situation of PAF's is better now.

The study of the role of SJVN Electric Power Projects in the Economic Development of Himachal Pradesh was also the subject matter of this Chapter (Chapter 3). Status of socio-economic development of Project Affected Families in comparison to those in Shimla and Kinnaur districts and the State average was examined in this chapter. It was concluded that socio-economic development status of Project Affected families was better not only in comparison to that of the Shimla and Kinnaur districts but in respect of the State average also.

In the matter of level of development, (where SJVN Project area was taken as a separate unit just like districts of Himachal Pradesh for the sake of meaningful analysis of level of development) the SJVN Project Affected Area was placed in high level of development in respect of sex ratio, literacy percentage, female literacy percentage, literacy percentage among SC/ST population, female literacy percentage among SC/ST female population, percentage of main workers to total population, percentage of female main workers to total female population and percentage of families placed above the poverty line.

Measurement of Development Index in respect of all the districts of Himachal Pradesh including SJVN Project Affected Area revealed that in terms of over-all socio-economic development the SJVN Project Affected was placed at the top position. It was realized that the efforts made by SJVN for the improvement of socio-economic conditions of the PAFs have been quite appreciable.

The SJVN has also initiated welfare measures for local people such as extension of educational and hospital facilities existing at the project area, construction of various link roads in the Project Area linking various villages to the National

Highway, construction of playground for Government High School at Village Gaura and additional rooms for the primary School at Shah, development of 5 Bigha land at Jhakri for Panchayat Ghar and Mahila Mandal Community Centre, construction of Play Ground for Middle School at village Sanarsa, construction of Helipad near Rampur, augmentation of existing government high school building and required furniture, etc. at Jhakri to open classes up to 10+2 standard, provision of a Mobile Health Van to provide health care services to the villages in the vicinity of project area in Shimla and Kinnaur Districts and assistance in construction of a 200 bedded Hospital at Rampur.

Furthermore, analysis of economic benefits to Himachal Pradesh made it clear that Nathpa-Jhakri Power Corporation has paid a dividend of Rs. 35.40 crores to Government of Himachal Pradesh for its operations for the year 2004-05. An interim dividend of Rs. 8.14 crores had already been paid by the NJPC to the State Government. From its nearly full year's commercial operations for the year 2004-05, the company has posted a profit of Rs. 298.43 crore on a turnover of Rs. 1098.28 crores. The company has declared a maiden dividend of Rs. 143.16 crore for the year 2004-05 to be shared by the equity partners, viz. Government of India and the Government of Himachal Pradesh in the ratio of 3:1 respectively. The Company's flagship venture 1500 MW Nathpa Jhakri Hydro-power Project had commenced commercial operations in October 2003 and has since been supplying valuable power to the northern grid beneficiary states viz. Delhi, Rajasthan, Uttar Pradesh, Punjab, Haryana, Himachal Pradesh, Jammu & Kashmir and Chandigarh. The Satluj Jal Vidyut Nigam (formerly NJPC) is presently executing three more projects in Himachal Pradesh, namely, 1020 MW Khab Hydroelectric Project, 412 MW Rampur Hydroelectric Project and 700 MW Luhri Hydroelectric Project.

Environmental concerns and role of hydropower projects in economic development have been discussed in Chapter 4. It is observed that economic development of a country and rise in standard of living of people largely depends upon the development and utilization of available natural resources. Water, the major natural resource, holds the key to economic prosperity and stability of the nation. Worldwide, Hydel power is the source to have been fully exploited by developed nations in order that they command a power generating source that is free from foreign control of sources of supply, and at the same time renewable.

River Valley Projects serve a basic necessity of a country whose economy is largely based on agriculture. The irrigation and power are crucial inputs for increased productivity. A high priority has, therefore, been given in the national planning process to the creation of the river valley projects ever since India became independent. The prominence given to the River Valley Projects in India's successive Five-Year Plans is consequent to the contribution to the prosperity of the country by the water resources projects.

Over the years, the information accumulated in course of working of river valley projects, revealed that the river valley projects, like all other development projects, have beneficial as well as adverse impacts. The two major criteria, that the project should maximize economic returns; and it should be technically feasible, are no longer considered adequate to decide the desirability or even the viability of the project. It is now widely recognized that the development effort may frequently produce not only sought for benefits, but either—often unanticipated—undesirable consequences as well, which may nullify the socio-economic benefits for which any project is basically designed. These impacts must be carefully assessed and balanced for achieving sustained benefits.

The construction of large hydropower projects in India and elsewhere is being criticized mainly on the grounds that it involves—Displacement of people from their traditional homes due to submergence caused by large storages created by hydropower projects; loss of traditional means of livelihood of the displaced people; change in the flow regime in the river channels downstream of dams/Hydel projects; and environmental and ecological effects of large storages created by the dams/Hydel projects

The construction of hydropower projects, large or small, does involve displacement of people from their traditional homes that calls for advance planning in acquisition of land, evacuation and rehabilitation of the effected people. The resettlement policy should be evolved on collaboration with the displaced people or their representatives by evolving a policy based on oustees' preference by—advance planning in acquisition of land; Provision for giving land for land; liberal compensation for submerged houses, land, trees, water mills, other affected properties; provision of free plots and some agriculture land for landless; and provision of social and other amenities such as drinking water, schools, healthcare centers, recreational facilities in oustees' colonies. It may also be noted that all type of developmental activities cause some type of displacement and other long-term effects and hydro projects are no exception.

It is very necessary for an agrarian country like ours to solve the problem of feeding the growing population and to build up its economy by development agriculture and industry on a sound footing to ensure availability of sustained power and abundant irrigation supplies. This can only be provided by large hydro projects. Environmental appraisal is an important responsibility and the function involves evaluation of environmental implications and incorporation of necessary safeguards for the activities having a bearing on environmental quality. The requirement is to consider environmental aspects as an integral part of a development project to achieve sustained development with minimum environmental degradation and prevention of long-term environmental side effects by incorporating mitigative measures. It is imperative to analyse whether the adoption of environmental measures is going to result in any short or long-term social and economic benefits or not. The object of environmental impact assessment is to ensure that development proceeds hand in hand with ecological preservation so as to achieve sustained growth. Environmental protection for sentimental reasons alone is neither desirable nor can it be defended.

Positive impacts of water resources projects outweigh the feared negative impacts and can be kept the minimum by integrating environmental concerns in the project planning right from the beginning. Large hydro projects have substantial positive environmental impacts, e.g. assured supply of drinking water and water for irrigation and industry, flood moderation, reduction in drought frequency, improved agricultural produce, production of hydropower, tourism, pisciculture, ground water recharge, development of industry and consequent economic progress and creation of additional employment opportunity, etc. Most of the projects under execution and projects planned for execution are run-of-the-river type development in Himachal Pradesh. Most of these projects involve a diversion structure in the form of a barrage or a trench weir, tunnel for water conductor system and underground power house. Thus, surface structures in many of the projects are mainly a barrage or weir and tail race outlet structure. In view of most of the projects being located underground, there is least environmental impact.

In order to minimize impact during construction, proper dumping areas with retaining structures and beams to reduce surface run-off along the excavates slopes, creation of setting structures for collection of waste oil, wetting of un-metalled roads regularly to prevent generation of dust, are strictly enforced. Maintenance of vehicles and stationary plants and machinery is carried on regular basis at shorter intervals to minimize the amounts of hydrocarbons contributed to atmosphere. For construction of underground structures, ventilation systems are installed to ensure adequate supply of oxygen inside and to throwout noxious fumes and dust generated due to blast.

Wherever possible, fish passage is being provided especially in the projects located on the main rivers, so that migratory fish can access normal spawning areas upstream of diversion structure of a hydropower project. Recently, fish ladder has been provided in Larji Hydroelectric Project (126 MW), nearing completion on the river Beas in Himachal Pradesh. In projects having reservoirs, adequate facilities are being developed. Developers also pay reasonable sum depending on the capacity of the Project to the Department of Fisheries, Government of Himachal Pradesh for the development of fish hatcheries in the project area or in its vicinity.

Environmental flow, i.e. release of water to the river downstream from the diversion structure is mandatory. It is now being ensured that river reach between the dam and the power station has sufficient water flows so that fish and other aquatic organism are able to reside in the affected reach. Simple studies for maintenance of ecosystem downstream of the diversion structure are being conducted. These studies besides, consider amount of water required for villages located in the bypass reaches for irrigation, domestic and other uses.

Most of the projects being executed are run-of-the-river schemes with simple diversion structures. There are no major storage reservoirs in many of the schemes which are under execution or have been planned for future. At places where there is inundation, mostly confined within the river banks, vegetation cover is cleared before reservoir filling, in order to prevent generation of greenhouse gases and disease vectors. There are no big reservoirs. Almost all the rivers and streams flow in step gradient with boulders in between or towards the banks, thus, there is no lack of dissolved oxygen in the river water downstream of the power house.

Social impact including loss of homes and agricultural areas are compensated as per resettlement and rehabilitation policy of the Government of Himachal Pradesh. It is being ensured that while executing hydroelectric projects in the state, people affected by the development of these projects are better-off than before the construction of the Project.

Loss of forests and habitats for wildlife are compensated mainly by carrying out compensatory afforestation in the catchments area as per Catchments Area Treatment Plan for the river basin. Mitigation measures are incorporated in the design of power plants.

Himachal Pradesh has hydroelectric power potential as one of its main resources. Initially most of the projects in the state were executed in the central sector. State was given a token share of the energy generated from these projects. Now, a new Hydel policy has been framed by the government of Himachal Pradesh. According to the policy, projects allotted to the private sector, at present 12 per cent free power is given to the state for the first twelve years and 18 per cent for the next eight years. After 40 years project is to be transferred to the state free of cost. Proper implementation of such policy would definitely benefit the economy of Himachal Pradesh not only in short-run but also in long-run.

Chapter 5 highlights the improvements in the standard of living as well as socio-economic conditions of the people in the state of Himachal Pradesh due to rapid exploitation of hydroelectric potential, available in the state mainly on the run-of-the-river which obviously proved most viable and economical. Most of these projects involve a diversion structure in the form of a barrage or a trench weir, tunnel for water conductor system and underground power house. In view of most of the projects being located underground, there is least environmental impact.

Himachal Pradesh has been averaging an annual growth rate of around 7 per cent since 1994-95, a commendable achievement, generally, at rates above the national gross domestic product (GDP) growth. With a per capita income of Rs. 30,138 in 2005-06, it ranks seventh among the states, with a rate of growth of per capita income higher than the all-India trend. The state has one of the lowest levels of poverty in the country with a mere 7.6 per cent of its population living below poverty line in 1998-99, compared to the all-India average of 26.2 per cent. The economy is largely depend on its service sector (which included electricity), which contributes 39.6 per cent to total income, though the manufacturing sector has also be vibrant as it makes up 34.5 per cent of the state income. As with other states, over time its economy has shifted from being more dependent on agriculture sector to focusing on industry and the service sector.

Despite the fact that cultivable area is just 14.4 per cent of the total area, 21 per cent of the state's income comes from agriculture. The forest cover of more than 65 per cent of the area yields returns worth about 4 per cent of the gross state domestic product. It is worth mentioning that majority of the labour force, 67.3 per cent, is engaged in the primary sector. The growth rate of agricultural production has increased from 0.8 per cent during the period 1994-95 to 1999-2000 to 7.4 per cent during 2001-02 to 2004-05. The fluctuations witnessed in the nineties have been warded-off since 2002 by laying emphasis on production of off-season vegetables, potatoes, pulses and oilseeds. Supply of high yielding variety seeds for maize, paddy and wheat helped improve food grain production. The state is one of the largest producers of horticulture products in the country and the production of fruits here was high at 0.56 million tonnes in 2003-04, with major proportion accounted by apple production.

The industrial sector has performed well, getting a boost with the removal of industrial licensing in the early nineties. The promulgation of special incentive schemes for investment led to many industries seeking a base in the State. Notwithstanding, the growth rate in manufacturing has tempered from an annualized average rate of 13.8 per cent during the period 1995-96 to 1999-2000 to 8.4 per cent during 2000-01 to 2004-05, the state is still in the top quartile in terms of gross output per capita and value added in the industries. Registered manufacturing accounts for more than 75 per cent of the total manufacturing with some of the key industries, being iron foundries, resin and turpentine, breweries, fertilizers and electronics. The weaving of wool garments and sericulture are key small-scale activities.

The service sector which has led the growth in other states has not picked up to the same extent this decade. The rate of growth of service sector at 9.8 per cent during 1995-96 to 1999-2000 was well above the national growth rate but has since slipped down to 5.4 per cent for the period 2001-02 to 2004-05. Tourism is one of the main sectors of the state with a compound annual growth of 17.5 per cent during 1998-2003. However, the state has not managed to attract the IT firms in large numbers. Given the good spread of infrastructure, it appears that the potential has not been exploited to the maximum for the service sector.

Himachal Pradesh ranks high on infrastructure provision in the country as it is the first state that has achieved close to 100 per cent electricity connections for its households. Given the topography of the state, the fact that 90 per cent of its population resides in rural areas and the fact that these 20,000 odd villages are scattered, this is a remarkable achievement. The state is among the few states with zero power deficits. It has a huge potential of hydropower generation at 2,03,086 MW (24 per cent of the total potential in the country). So far, 3,942 MW of hydroelectric power has been harnessed, of which only 326.8 MW is under public sector. The state is keen to encourage private sector participation in the power sector.

In the matter of communication facilities, the state has also provided well, with a telephone density of about eight relatives to five for the country as a whole. The state has recorded highest growth of mobile phone users in 2002-03 and the penetration of mobile phones is more than that in neighbouring states Haryana and Uttarakhand. The number of post offices per lakh of population is highest here at 44 per lakh of population, thereby connecting remote villages as well to the outside world. The state has a well developed banking sector—the number of bank branches per lakh of population is more than 12 as compared to seven for the country as a whole, this is the second highest among the states.

The employment scenario in the state is better than in other parts of the country—with non-workers forming 40 per cent of the total population, well below the 60 per cent seen at the national level, it appears that economic growth has spread across sections of the society. The work participation rate for females ranks among the highest in the country at 47.8 per cent, higher than Punjab's 34.2 per cent and Haryana's 39.5 per cent.

Himachal Pradesh is shining as a model state in India in achieving developmental goals. The drop in the fertility rate from 3 to 2.1 during the period 1992-93 to 1998-99 has eased the population pressure on the state as far as provision of public services to all are concerned. The state gives prime importance to health and education. In 2003-04, it allocated 6 per cent of its disbursements to education, increasing this to 11.5 per cent in 2005-06, compared to the median state's 8.1 per cent. Thus the expenditure per child in the age group 6-14 years was Rs. 9,465 in 2003-04, while the median state spent only half this amount per child. In 2002, the dropout ratio from primary school was a meager 9.7 percent, compared to Punjab's 32.8 per cent and Haryana's 12.4 per cent. Again, this achievement has to be viewed in the light of the difficult terrain and issues of accessibility to every settlement. It is true that many schools are still single teacher schools but in a recent survey done by NGO Pratham all across the country, the students in Himachal schools fared much better than those in the neighbouring states when tested for reading and arithmetic skills. With five technical colleges, seven medical colleges and 69 colleges for arts, science and commerce, Himachal Pradesh also provides well for higher education facilities.

The health infrastructure in this small state includes 50 civil hospitals, 60 community health centres, 438 primary health centres and 2,067 sub-centres. The 52.1 per cent of the births in the state were assisted by technical personnel in 2002-03, compared to the national median of 42.8 per cent. With 79 per cent of the children in the age group 12-35 months being given full immunization, the state is just behind Tamil Nadu and Kerala where the levels of immunization are the highest. However, the infant mortality rate of 51 per 1,000 births needs to be improved upon, as it is higher than the median state's 42.

On the fiscal front, Himachal Pradesh has a high gross fiscal deficit to GSDP ratio of 6.9 per cent in 2004-05. This is, however, half of the average of 12.7 per cent in the period 2001-04 and the progress towards the target of 3 per cent set by the Twelfth Finance Commission is welcome. It must be remembered that Himachal Pradesh is one of the 11 special category states, along with the seven states in the northeast and the hill states of Sikkim, Uttarakhand and Jammu and Kashmir, all of which are perceived to have special needs for development. Though Himachal Pradesh has been doing well at providing basic amenities and income opportunities to its populace, for growth to be sustained it will be crucial that revenue collection be stepped up. At 4.9 per cent of the GSDP, own tax revenue is still below comparator state Uttarakhand's 6.9 per cent. Social sector expenditure and capital outlays have reduced over time as a percentage of GSDP and are less than that of other special category states. However, a worrisome feature is the high debt-GSDP ratio of 73.7 per cent in 2004-05.

Policy Implications

The only strategy needed for exploitation of vast hydroelectric power resources is to produce as much energy as possible, as fast as possible, with minimum cost and with minimum environmental negative consequences for placing the economy of Himachal Pradesh on a faster track of development. It is quite clear that hydel power resource development will make the state economy self-dependent if the hydel power potential is harnessed in the shortest possible time. Due to financial constraints, it has become difficult for the state government to tap enormous power potential without financial support from non-government organizations. It is realized that giving opportunity to private sector for harnessing power potential is not showing optimum economic benefits to the state because the objective of the private sector is to earn only profit and they are least concerned with the development of the state.

In order to increase the number of hydroelectric projects and ultimately power generation, efforts should be made by the state government to enhance the power generation capacity of joint venture corporations already in existence in the State of Himachal Pradesh. Further, an attempt to form joint ventures with neighbouring states' hydroelectric power projects will deliver economic benefits not only to Himachal Pradesh, but also to the particular neighbouring states.

While exploiting hydropower resources, we have to keep in mind that it is better to walk in the right direction than to run in the wrong one. As per conditions laid down in the agreement with private investors for undertaking hydropower mini-micro and major projects the site is to be handed over for operation only for a period of 40 years after commercial operation and the project is to be transferred to the Government of Himachal Pradesh free of cost after expiry of agreement period. If this condition is implemented judiciously, it will give a sound footing in the long-run for the economy of this State.

Further, there is a need to encourage cooperative sector by reserving economically viable mini-micro sites which will involve local community and increase the employment opportunities to local people. Though this feature of involvement of cooperative sector has already been started by the government, yet sincere effort in this direction has not been made. Therefore, a blueprint should be prepared for roping in of cooperative sector. This type of attempt will be proved fruitful for the economic prosperity of the state in particular and country in general.

APPENDIX 1

Abstract of National Policy on Resettlement and Rehabilitation for Project Affected Families-2003 (NPRR-2003): Ministry of Rural Development (Department of Land Resources). Resolution: New Delhi, the 17th February, 2004* .

F. No. Acq. 1301/4/2003-LRD-WHEREAS the Government of India, Ministry of Rural Development (Gramin Mantralaya), Department of Land Resources (Bhumi Sansadhan Vibhag), has formulated a National Policy on Resettlement and Rehabilitation for Project Affected Families-2003;

AND whereas the Government of India desires that the contents of the said Policy be brought to the notice of the general public and given widest publicity;

NOW, therefore, it is directed that the National Policy given in the Schedule hereto annexed be published in the Gazette of India, Extraordinary Part-I, Section 1, dated 17th February, 2004.

SCHEDULE

Subject: National Policy on Resettlement and Rehabilitation of Project Affected Families-2003.

CHAPTER 1

POLICY

1. Preamble

1.1 Compulsory acquisition of land for public purpose

* Taken from Social Action, Vol. 55, April-June, 2005, pp. 179-98.

including infrastructure projects displaces people, forcing them for given up their home, asset and means of livelihood. Apart from depriving them of their lands, livelihood and resource base, displacement has other traumatic psychological and socio-cultural consequences. The Government of India recognizes the need to minimize large scale displacement to the extent possible and, where displacement is inevitable, the need to handle with utmost care and forethought issues relating to R & R of Project affected families. Such an approach is especially necessary in respect of tribal, small and marginal farmers and women.

1.2 The system of extending cash compensation does not, by itself in most cases, enable the affected families to obtain cultivable agricultural land, homestead and other resources which they have to surrender to the State. The difficulties are most acute for persons who are critically dependent on the acquired assets for their subsistence/livelihoods, such as landless agricultural workers forest dwellers, tenants and artisan, as their distress and destitutions is more severe, and, yet they are are not eligible for cash compensation.

1.3 Some states and Central Ministries/Departments have their own Policies and Guidelines for Resettlement and Rehabilitation. However, a National Policy on R&R of Project Affected Families has not so far been enunciated. This Document aims at laying down basic norms and packages in the shape of a Policy which would, henceforth be referred to as the National Policy on the R & R of PAF-2003 (NPRR-2003).

1.4 The Policy essentially addresses the need to provide succor to the asses less rural poor, support the rehabilitation effort of the resources poor sections, namely, small and marginal farmers, SCs/STs and women who have been displaced. Besides, it seeks to provide a broad canvas for an effective dialogue between the PAF and Administration for R & R. Such a dialogue is expected to enable timely completion of projects with a sense of definiteness as regard cost and adequate attention to the needs of displaced person especially the resource poor sections. The intentions is to impart greater flexibility for interaction and negotiation so that the resultant Package gains all round acceptability in the shape of a workable instrument providing satisfaction to all stakeholders/ requiring bodies.

1.5 The National Policy on the R & R of PAF will be in the form of broad guideline and executive instructions for guidance of all concerned and will be applicable to Projects displacing 500 families or more enmesh in plain area and 250 families nemeses in hilly area, Desert Development Programme blocks, area mentioned Schedule V and Schedule VI of the Constitution of India. It is expected that the appropriate Government and Administrator for R & R shall implement this policy in letter and spirit in order to ensure that the benefits envisaged under the Policy reaches the Project Affected Families, especially resource poor section including SCs/STs.

1.6 The rehabilitation grants and other monetary benefits proposed the Policy would be minimum and applicable to all project affected families whether belonging to BPL or non-BPL families States where R& R packages are higher than proposed in the Policy are free to adopt their own packages.

CHAPTER II

2. Objectives of the Policy

2.1 The objectives of the Policy are as follows:

(a) To minimize displacement and to identify non-displacing or least-displacing alternatives,

(b) To plan the R & R of PAFs including special needs of Tribals and vulnerable sections,

(c) To provide better standard of living to PAFs, and

(d) To facilitate harmonious relationship between the Requiring Body and PAFs through mutual cooperation.

CHAPTER III

3. Definitions

3.1 The definition of various terms used in this Policy Document is as follows:

(a) Administrator for Resettlement and Rehabilitation means an officer not below the rank of District Collector of the State Government appointed by it for the purpose of resettlement and rehabilitation of the PAF of the Project concerned provided that if the appropriate Government in respect of the project is the Central Government, such appointment shall be made in consultation with the Central Government;

(b) "affected zone", in relation to a project, means declaration under para 5.1 of this Policy by appropriate Government area of villages or locality under a project for which the land is being acquired under Land Acquisition Act, 1894 or any other Act in force or an area that comes under submergence due to impounding of water in the reservoir of the projects;

(c) "agricultural family" means a family whose primary mode of livelihood is agriculture and includes family of owner as well as sub-tenants of agricultural land, agricultural laborers, occupiers of forest lands and of collectors of minor forest produces;

(d) "agricultural labourer" means a person normally resident in the affected zone for a period of not less that three years immediately before the declaration of the affected zone who does not hold any land in the affected zone but who earns his livelihood principally a manual labor on agricultural land therein immediately before such declaration and who has been deprived of his livelihood;

(e) "agricultural land" includes land used or capable of being used for purpose of:

(a) agriculture or horticulture;
(b) dairy farming, poultry farming, pisciculture, breeding or livestock and nursery growing medical herbs;
(c) raising of crops, grass or garden produce and
(d) land used by an agriculturist for the grazing of cattle, but does not\include land used for cutting of wood only;

(f) Appropriate Government means:

(i) In relation to a question of land for the purpose of the Union, the Central Government;
(ii) In relation to project which is executed by Central Government agency/Central Government undertaking or by any other agency on the orders/directions of Central Government, the Central Government, otherwise the State Government; and
(iii) In relation to acquisition of land for other purpose, the State government.

(g) 'BPL Family' the below Poverty Line Families shall be those as defined by the Planning Commission of India from time to time.

(h) Commissioner for R & R, in relation to a project means, the Commissioner for R & R appointed by the State Government not below the rank of Commissioner/Secretary of the Government.

(a) Displaced family means any tenure holder, and Government less or owner of other, property, who on account of acquisition of his land including plot in *abadi* or other property in the affected zone for the purpose of the project has been displaced from such land or other property;

(i) "family" means project affected family consisting of such person, his or her spouse, minor sons, unmarried daughters, minor brothers or unmarried sisters, father/mother and other members residing with him and dependent on him for their livelihood;
(j) "holding" means the total land held by a person as an occupant or tenant or as both;
(k) "marginal farmer" means a cultivator with an un-irrigated land holding up to one hectare or irrigated land holding up to half hectare;
(l) "non-agricultural labourer" means a person who is not an agricultural laborer but is normally residing in the affected zone for a period of not less than three years immediately before the declaration of the affected zone and who does not hold any land under the affected zone but who earns his livelihood principally by manual labor or as a rural artisan immediately before such declaration and who has been deprived earning his livelihood principally by manual labour or as such artisan in the affected zone;
(m) "notification" means a notification published in the official Gazette;
(n) "Occupiers" mean members of Scheduled Tribe community in possession of forest land prior to 25th October, 1980;
(o) "Project" means a project displacing 500 families or more enmasse in plain area and 250 families or more enmasse in hilly area, DDP blocks, areas mentioned in Schedule V and Schedule VI of the Constitution of India as a result of acquisition of land for any project;

(p) "Project affected family" means a family person whose place of residence or other properties or source of livelihood are substantially affected by the process of acquisition of land for the project and who has been residing continuously for a period of the affected zone of practicing any trade, occupation or vocation continuously for a period of not less that three years in the affected zone, preceding the date of declaration of the affected zone;

(q) "Resettlement zone", in relation to a project means the declaration of any area under Para 5.12 of this Policy by the appropriate Government acquired or proposed to be acquired for resettlement and rehabilitation of Project Affected Families in a resettlement zone.

(r) "Requiring Body" shall means any company, a body corporate an institution, or any other organization for whom land is to be acquired by the appropriate government and includes the appropriates Government if the acquisition of land is for such Government either for its own use of for subsequent allotment of such land in public in the interest to a body corporate, institution or any other organization or to any company under lease, license or through any other system of transfer of land to such company, as the case may be;

(s) "Small farmer" means a cultivator with an un-irrigated land holding up to two hectares or with an irrigated land holding up to one hectare.

CHAPTER IV

4. Appointment of Administrator and Commissioner for Resettlement and Rehabilitation and their Powers and Functions

4.1 Where the appropriate Government is satisfied that acquisition of land for any project involves displacement of 500 families or more enmasse in plain area and 250 families or more enmasse in hilly area, DDP blocks, areas mentioned in Schedule V and Schedule VI of the Constitution of India as a result of acquisition of land for any project, it shall, by notification/appoint in respect of that project, an officer not below the rank of District Collector of the State Government to be the Administrator for R & R in respect of that project:

Provided, that if the appropriate Government in respect of the project is the Central Government, such appointment shall be made in consultation with the Central Government.

4.2 The Administrator for R & R shall be assisted by such officers and employees as the appropriate Government may provide.

Administrator for Resettlement and Rehabilitation

4.3 Subject to the superintendence, directions and control of appropriate Government and Commissioner for R & R the Administrator for R & R shall take all measures for the rehabilitation and resettlement of all project affected families in respect of that project.

4.4 The overall control and superintendence of the formulation of R & R plan and execution of the same shall vest in the Administrator, R & R.

Powers and Functions of Administrator

4.5 Subjects to any general or special order of appropriate Government, the Administrator for R & R shall perform the following functions/duties;

(i) Minimize displacement of person and identify non-displacing or least displacing alternative in consultation with the requiring body;

(ii) hold consultation with the project affected families while preparing a resettlement and rehabilitations scheme/plan;

(iii) ensure that interests of the adversely project affected families of scheduled tribes and weaker sections are protected;

(iv) prepare a draft plan/schemes of R & R as required under Chapter V of this policy;

(v) prepare a budget including estimated expenditure of various components of acquisition of land, R & R activities or programmes in consultation with representatives of the PAFs and requiring body for whom the land is acquired;

(vi) adequate land for the project and also for setting the project affected families;

(vii) allot land and sanction benefits to PAFs; and

(viii) Perform such other functions as the appropriate Government may from time to time by order in writing assign.

Delegation of powers of Administrator

4.6 Administrator for R & R may be order in writing, delegate such of administrative power conferred and duties imposed on him by or under this Policy to any officers not below the rank of Tehsildar or equivalent.

4.7 All officers and staffs appointed by the appropriate Government under this Policy shall be subordinate to the Administrator for R & R.

Commissioner for R & R

4.8 The State Government shall appoint an officer of the rank of Commissioner/Secretary of that Government of resettlement and rehabilitation in respect of such projects to which this Policy applies to be called the Commissioner for R & R.

4.9 For the purpose of this Policy, the Administrator for R & R and other officers and employees appointed for the purpose for R & R of PAF shall be subordinate to the Commissioner for R & R.

Functions of Commissioner for R & R

4.10. The commissioner shall be responsible for supervising the formulation of R & R plans/schemes, proper implementation of such plans/schemes and redressed for grievance as mentioned in Chapter VII of this Policy.

CHAPTER V

5. Schemes/Plans for Resettlement and Rehabilitation

The procedure mentioned in this Chapter shall be followed for declaration of Affected Zone, carrying out survey and census of PAF, Assessment of Government land available and land to be acquired for the purpose of R & R, preparation of draft scheme/plant for R & R and its final publications.

Declaration of Affected Zone

5.1 The appropriate Government may if it is of the opinion that acquisition of land for a project is likely to displace 500 families or more enmasse in pain areas and 250 families or more, in hilly areas, DDP blocks, areas mentioned in Schedule V and Schedule VI of the constitution of India declared by notification in the Official Gazette, area of villages or localities as an affected zone of the project and thereupon the contents of this Policy shall apply to the project involved.

Procedure to be followed for survey and census of PAFs etc.

5.2 Every declaration made under Para 5.1 of the Policy shall be published in at least two daily newspapers, one of them should be in the local vernacular having circulation in villages or area which are likely to be affected and also by affixing a copy of the notification on the Notice Board of the concerned *Gram Panchayats* and other prominent place or places in the affected zone.

5.3 Once the declaration is made under para 5.1 of the Policy, the Administrator for R & R shall undertake a survey for identification of the persons and their families likely to be affected by the project.

5.4 Every survey shall contain the following village-wise information of the project affected families.

(i) Members of families who are permanently residing, practicing any trade, occupation or vocation in the project affected area.

(ii) Project Affected Families who are likely to loose the house, agriculture land, employment or are, alienated wholly or substantially from the main source of their trade occupation or vocation.

(iii) Agricultural laborer and non-agriculture laborers.

(iv) Project affected families who are having possession of forest lands prior to 25th October, 1980, that is, prior to the commencement of the Forest (Conservation) Act, 1980.

5.5 Every survey undertaken under Para 5.4 shall be completed within a period of ninety days from the date of declaration made under para 6.1.

5.6 On the expiry of the period of ninety days as aforesaid, the Administrator for R & R shall, by notification and in such other manner so as to reach all person likely to be affected, publish a draft of the details of the findings of the survey conducted by him for inviting objections and suggestions all persons likely to be affected thereby.

5.7 On the expiry of thirty days from the dates of publication of the draft of the details of survey and after considering the objections and suggestions received by him in this behalf, the administrator for R & R shall submit the final details of survey with his recommendations to the State Government.

5.8 Within forty-five days from the date of receipt of the recommendations of the Administrator for R & R, the State Government shall publish the final details of survey in the official Gazette.

5.9 The Administrator for R & R shall ensure that the PAF may be settled preferably in group or groups and such sites should form a part of existing *gram panchayat* as far as possible;. However, it has to be ensured that PAF's may be resettled with the host community on the basis of equality and mutual understanding, consistent with the desire of each group to preserve its own identity and culture.

Assessment of land available for R & R

5.10 For the purposes of para 5.9 above, the Administrator for R & R shall draw up a list of lands which may be available in any existing Gram Panchayat for R & R of PAFs.

5.11 The lands drawn up under para 5.10 shall consists of:

(a) Government waste lands and any other land vesting in the Government available for allotment to project affected families.

(b) If sufficient government land is not available there, then land to be acquired for the purposes of R & R schemes. However, the administrator for R & R should ensure that such acquisition of land should not lead to another list of affected families.

Declaration of Resettlement Zone

5.12 The appropriate Government shall by notification declare any area acquired for proposed to be acquired for resettlement and rehabilitation of project affected families as a resettlement zone.

Power to acquire land for R & R

5.13 The Administrator for R & R on behalf of the

appropriate government, may either compulsorily acquire keeping in view the contents of Para 5.11 (b) above any land under the Land Acquisition Act, 1894, or purchase land from any person through consent award and may enter into an agreement for this purpose

Draft Scheme/Plan for R & R

5.14 After completion of base line survey and census of Project Affected Families and assessment of requirement of land for resettlement as mentioned in Paras 5.3 and 5.11, the Administrator for R & R shall prepare a draft scheme/plan for the R & R for PAFs in consultation with representatives of PAFs including women, chairpersons of elected Panchayati Raj Institutions within which the project area is situated.

Management of Funds for R & R

5.15 While preparing a draft scheme/plan, the Administrator for R & R shall ensure that the cost of R & R scheme/plan should be an integral part of the cost of Project for which the land is being acquired and the entire expenditure of R & R benefits and other, expenditure for R & R of PAFs are to be borne by the requiring body for which the area is being acquired.

5.16 It shall be the responsibility of the requiring body to provide sufficient funds to the Administrator for R & R for proper implementation of R & R schemes/plans of PAFs.

5.17 The Administrator for R & R shall keep proper books of accounts and records of the funds placed at his disposal to submit periodical return to the Appropriate Government in this behalf.

5.18 Every draft scheme/plan of R & R prepared shall contain the following particulars namely:

(a) the extent of area to be acquired for the project and the name of the corresponding village;

(b) a village-wise list of project affected families and likely number of displaced persons, family-wise and the extent and nature of land and immovable property in their possession indicating the survey numbers thereof held by such persons in the affected zone;

(c) a list of agricultural laborer in such area and the names of such persons whose livelihood depend on agricultural activities;

(d) a list of persons who have lost or are likely to lose their employment or livelihood or who have been alienated wholly and substantially from their main sources of occupation or vocation consequent to the acquisition of land for the project;

(e) a list of occupiers, if any;

(f) a list of public utilities and Government buildings which were likely to be affected;

(g) a comprehensive list of benefit and packages which are to be provided to project affected families;

(h) details of the extent of land available which may be acquired in settlement area for resettling and allotting of land to the project affected families;

(i) details of basic amenities and infrastructure facilities which are to be provided for resettlement;

(j) the time schedule for shifting and resettling the displaced families in resettlement zone; and

(k) Such other particulars as the Administrator for R & R may think fit to include for the information of the displaced persons.

Final Publication of Schemes/Plan of R & R

5.19 The Administrator for R & R shall submits the drafts scheme/plan for R & R to the State Government of its approval. It will be responsibility of the State Government to obtain the consent of requiring body before approving the same. The draft scheme/plant may be published in the official Gazette to give wide publicity to the same in the affected zone.

5.20 Upon notification of a scheme/plan, the same shall come into force.

CHAPTER VI

6. R & R Benefits for Project Affected Families

6.1 The resettlement and rehabilitation benefits shall be extended to all the PAF's whether belonging to below poverty line or non-BPL.

6.2 Any project affected families owning house and whose housed has been acquired may be allotted free of cost house site to the extent of actual loss of area of the acquired house but not more than 1560 sq. m. of land in rural areas and 75 sq. meter of land in urban area

6.3 Each PAF of BPL category shall get a one time financial assistance of Rs. 25,000 for house construction. Non-BPL families shall not be entitled to receive this assistance.

6.4 Each PAF owning agricultural land in the affected zone and whose entire land has been acquired may be allotted agricultural land or cultivable waste land to the extent of actual land loss subject to a maximum of one hectare of irrigated land or two hectares of un-irrigated land/cultivable waste land subject of availability of Government land in the district.

6.5 Stamp duty and other fees payable for registration shall be borne by the requiring body.

6.6 The land allotted under para 6.4 shall be free from all encumbrances. The land allotted may be in the joint names of wife and husband of PAF.

6.7 In case of allotment of wasteland/degraded land, in lieu of acquired land, each PAF shall get financial assistance of Rs. 10,000 per hectare for land development. In case of allotment of agriculture land, a one time financial assistance of Rs. 5000 per PAF for agricultural production shall be given.

6.8 Each PAF having cattle shall get financial assistance of Rs. 3000 for construction of cattle shed.

6.9 Each PAF shall get financial assistance of Rs. 5000—as transportation cost for shifting of building material, belonging and cattle etc. from the affected zone to the resettlement zone.

6.10 Each PAF comprising of rural artisan/small trader and self employed person shall get one time financial assistance of Rs. 10,000 for construction of working shed/shop.

6.11 Each PAF owning agricultural land in the affected zone and whose entire land has been acquired shall get one time financial assistance equivalent to 750 days minimum agricultural wages for loss of livelihood where neither agricultural land nor regular employment to one member of the PAF has been provided.

6.12 Each PAF owning agricultural land in the affected zone and whose entire land has not been acquired and consequently he becomes a marginal farmer shall get one time financial assistance equivalent to 500 days minimum agricultural wages.

6.13 Each PAF owning agricultural land in the affected zone and who consequently becomes a small farmer shall get one time financial assistance equivalent to 375 days minimum agricultural wages.

6.14 Each PAF belongings to the category of agricultural laborer or non-agricultural laborer shall be provided a one time financial assistance equivalent to 625 days of the minimum agricultural wages.

6.15 Each displaced PAF shall get a monthly subsistence allowance equivalent to 20 days of minimum agricultural wages per month for period of one year upto 250 days of MAW.

6.16 In case of acquisition of land in emergent situation such as under Section 17 of the Land Acquisition Act, 1894 or similar provision of other Act in force, each PAF shall be provided with transit accommodation, pending resettlement and rehabilitation scheme. Such families shall also get R & R benefits as mentioned in above para under the policy.

6.17 Acquisition of Long Stretches of Land: In case of projects relating to Railway Line, Highways, Transmission Lines and laying pipelines wherein only a narrow stretch of land extending over several kilometers is being acquired, the Project Affected Families will be offered an *ex-gratia* amount of Rs. 10,000 per family, and no other R & R benefits, shall be available to them.

6.18 The PAF shall be provided necessary training facilities for development of entrepreneurship to take up self-employment projects at the resettlement zone as part of R & R benefits.

6.19 The Project Affected Families who were in possession of forest lands prior to 25th October, 1980 shall get all the benefits of R & R as given in above paras under the Policy.

6.20 The PAFs of Scheduled Caste category enjoying reservation benefits in the affected zone shall be entitled to get the reservation benefits at the resettlement zone.

6.21 R & R Benefits for Project Affected Families of Scheduled Tribes:

6.21.1 Each PAF of ST category shall be given preference in allotment of land.

6.21.2 Each tribal PAF shall be entitled to get R & R benefits mentioned in above Paras under the Policy.

6.21.3 Each Tribal PAF shall get additional financial assistance equivalent to 500 days minimum agriculture wages for loss of customary right/usages of forest produce.

6.21.4 Tribal PAFs shall get land free of cost for community religious gathering.

6.21.5 Tribal PAFs resettled out of the district talk and will get 25% higher R & R benefits in monetary terms.

6.21.6 The Tribal Land alienated in violation of the laws and regulations in force on the subject would be treated as null and void and the R & R benefits would be available only to the original tribal and owner.

6.21.7 The Tribal families residing in the PAF areas having fishing rights in the river/ponds/ dam shall be given fishing rights in the reservoir area.

6.21.8 Tribal PAFs enjoying reservation benefits in the affected zone VC shall be entitled to get the reservation benefits at the resettlement zone.

6.22 Basic Amenities To Be Provided At Resettlement Zone.

6.22.1 While shifting population of the affected zone to the resettlement zone, the administrator for R & R may as far as possible ensure that:

(a) In case the entire population of the village area to be shifted belongs to particular community, such population/families may be resettle in a compact area so that socio-cultural relations among shifted families are not disturbed.

(b) In case of resettlement of scheduled caste PAFs it may be ensured that they are resettled in site close to the villages.

6.22.2 The PAF shall be provided the basic amenities and infrastructural facilities at the resettlement sites as per norms specified by the Appropriate Government. It is desirable that provision of drinking water, electricity, schools, dispensaries and access to the resettlement site among other be included in the re-settlement plan formulated by the Administrator for R & R.

CHAPTER VII

7. Dispute Redressal Mechanism

7.1 R & R Committee at Project Level

7.1.1 In respect of every project to which this policy applies, the state Government shall constitute a committee under the chairmanship of the Administrator of that project to be called as the Resettlement and Rehabilitation Committee to monitor and review the progress of implementation of scheme/plan of R & R of the PAFs.

7.1.2 The R & R Committees constituted as above shall *inter-alia* include as one of its members—

(a) a representative of women residing in the affected zone,
(b) a representative each of scheduled caste and scheduled tribes residing in affected zone,
(c) a representative of a voluntary organizations
(d) a representative of lead bank,
(e) chairman of his nominee of the PRIs located in the affected zone, and
(f) MPS/MLAs of the area included in the affected zone.

7.1.3 Procedure regulating the business of the R & R committee, its meeting and other matter connected thereto shall be prescribed by the Appropriate Government.

7.2 Grievance Redressal Cell

7.2.1 In respect of ever/projects to which this policy applies, the State government shall constitute a Grievance Redressal Cell under the Chairmanship of the Commissioner for R & R for redressal of grievances of PAFs.

7.2.2 The composition, powers, functions and other matters relating to the functioning of the Grievance Redressal Cell shall be such as may be prescribed by the appropriate Government.

7.2.3 Any PAFs if aggrieved, for not being offered the admissible R & R benefits as provided under this policy, may move an appropriate petition for redressal of his grievances to the Grievance Redressal Cell.

7.2.4 The forms and manner in which and the time within which complaints may be made to the Grievance Redressal Cell and disposed of shall be such as may be prescribed by the appropriate government.

7.2.5 The grievance redressal cell shall have the power to consider and dispose of all complaints relating to R & R against the decision of the Administrator/R & R Committee at Project level for R & R and issue such directions to the Administrator for R & R as it may deem proper for the Redressal of such grievances.

7.3 Inter-State Projects

7.3.1 In case a project covers an area in more than one State or States or a Union territory where the project affected families are or had been residing, or proposed to be resettled, the Central Government in the Ministry of Rural Development (Department of Land Resources) shall in consultation with concerned States or Union territory, as the case may be appoint the Administrator for R & R and the Commissioner for R & R for the purposes of this Policy.

7.3.2 The method of implementation of plans/schemes for R & R shall be mutually discussed by the State Government and the Union territory administration and the common plan shall be notified by the Administrator for R & R in the State or Union territory administration as agreed to, in accordance with the procedure laid down in this policy.

7.3.3 If any difficulty arises in the implementation of the schemes/plans, the matter shall be referred to the Central Government in the Ministry of Rural

Development (Department of Land Resources) for its decision and the decision of the Central Government shall be binding on the concerned States and union territory.

CHAPTER VIII

8. National Monitoring Committee

8.1 The Central Government, Ministry of Rural Development, Department of Land Resource shall constitute a National Monitoring Committee, to be chaired by the Secretary, Department of Land Resources for reviewing and monitoring the progress of implementation of R & R schemes/plan relating to all projects to which this policy applies. The committee will have the following or his nominees not below the rank of Joint secretary as its members; secretary, Planning Commission, Secretary, M/O Social Justice and Empowerment, Secretary, M/O Water Resources; Secretary, M/O Tribal Affairs; Secretary, M/O Railways; Secretary, M/O Power; and Secretary, M/O coal.
Besides The Secretary of the Administrative Ministry/ Department of the project for which the land is to be acquired shall be invited as one of the Members. The functions and duties of this committee shall be prescribed by this ministry.

8.2 The National Monitoring Committee shall be serviced by the National Monitoring Cell to be constituted by the Department of Land Resources for reviewing and monitoring the progress of implementation for R & R schemes/plan relating to all projects to which this Policy applies

National Monitoring Cell

8.3 National Monitoring Cell constituted under this Policy shall be headed by an officer not below the rank of Joint Secretary to the Government of India.

Applicability

8.4 The National Policy of Resettlement and Rehabilitation of Project Affected Families (NPRR-2003) shall come into effect from the date of its publication in the Gazette of India (Extra-ordinary).

APPENDIX II

Sale of Power

(in Million Kwh)

Year	*Sale within the State*	*Sale Outside the State*	*Total*
1974-75	191.8	94.7	286.5
1975-76	225.0	107.7	332.7
1980-81	264.73	147.13	411.86
1984-85	470.02	217.28	687.30
1985-86	563.32	223.93	787.25
1989-90	897.10	580.88	1477.98
1990-91	1008.74	901.90	1910.64
1991-92	1022.02	818.23	1840.25
1992-93	1083.28	823.35	1906.63
1993-94	1155.63	756.92	1912.55
1994-95	1339.68	984.65	2324.33
1995-96	1597.68	1049.10	2646.78
1996-97	1757.61	963.30	2720.91
1997-98	1946.54	954.21	2900.75
1998-99	2083.42	963.176	3046.596
1999-00	2181.741	938.947	3120.688
2000-01	2205.8	615.6	2821.4
2001-02	2331.837	548.837	2880.697
2002-03	2519.003	688.026	3207.029
2003-04	2726.324	1692.889	4419.213

Source : Statistical Outline of Himachal Pradesh (various issues), Directorate of Economics and Statistics, Shimla.

APPENDIX III

Power Consumption in Himachal Pradesh

(Million kWh)

Year	*Domestic*	*Commercial*	*Industrial*	*Government irrigation and Agriculture*	*Total*
1974-75	34.1	20.1	40.1	2.1	209.2
1975-76	42.4	24.0	47.7	2.6	20.5
1976-77	46.1	23.1	59.9	3.4	225.7
1977-78	50.0	23.9	40.9	5.4	202.4
1980-81	62.4	32.6	107.5	5.7	264.7
1981-82	70.5	34.9	130.5	6.5	285.9
1982-83	80.5	40.4	156.5	9.1	324.5
1983-84	92.5	45.5	207.0	11.9	804.0
1984-85	100.8	43.4	265.6	17.6	470.0
1989-90	196.6	73.6	443.5	11.3	897.1
1991-92	253.1	83.7	487.2	12.4	1022.0
1992-93	283.1	88.0	512.2	13.5	1083.3
1993-94	312.5	91.7	552.4	13.6	1155.6
1994-95	348.5	103.0	655.9	15.3	1339.6
1995-96	387.5	112.0	818.2	16.1	1597.6
1996-97	426.7	120.5	910.6	16.0	1757.6
1997-98	474.3	134.8	1019.6	17.3	1946.5
1998-99	537.5	139.8	1073.4	18.7	2083.4
1999-00	594.4	148.8	1111.4	19.1	2181.7
2000-01	636.5	161.6	1277.2	19.2	2205.8
2001-02	664.4	175.0	1324.8	18.0	2331.8
2002-03	712.9	187.7	1452.6	20.4	2516.5
2003-04	769.3	208.3	1338.0	19.4	2726.3

Source : Statistical Outline of Himachal Pradesh (various issues), Directorate of Economics and Statistics, Shimla.

APPENDIX IV

Status of Private Sector Hydroelectric Projects in Himachal Pradesh

Name of Project/capacity	*Executing agency*	*Date of signing of MOU/IA*	*Status of clearance*	*Status of project*
(1)	(2)	(3)	(4)	(5)
Allian Duhangan (192 MW)	M/s Rajasthan Spinning and Weaving Mills Ltd.	28.8.93/22.2.2001.	TEC— 20.8.2002 Forest—18.10.2002 MOEF—12.12.2000	The company is in process of achieving Financial Closure and commence most of work on the projct shortly.
Karcham Wangtoo (1000 MW)	M/s Jaypee Karcham Hydro Corporation Ltd.	28.8.93/18.11.99	TEC—31.3.2003 Site clearance—14.9.2001 MOEF—In Process Forest—21.11.2002 (In Principle)	The company is in the process of achieving remaining clearances and acquiring land for the project.
Patikari (16 MW)	M/s East India Petroleum Ltd./M/s Patikari Power Pvt. Ltd.	21.6.2000/ 9.11.2001	TEC—27.9.2001 Forest—1.11.2004 MOEF—in process Financial Closure—in process	The company has commenced the work in January 2005.
Sainj (3 MW)	M/s East Indian Petroleum Ltd.	21.6.2000/ 24.7.2000	TEC—23.4.2000/ Forest in process MOE—In process	The company is in process of obtaining remaining clearances for the project.

(Contd.)

APPENDIX IV (*Contd.*)

(1)	*(2)*	*(3)*	*(4)*	*(5)*	*(6)*	*(7)*	*(8)*
Malana-II (100 MW)	M/s Everest Power Pvt. Ltd.		27.5.2002	TEC—15.10.04 MOEF—in process Financial Closure—in process		The company is in process of obtaining other clearances for the project	
Dhiulasidh (40 MW)	M/s GVK Industries Ltd. in HPSEC for TEC		20.6.2002			DPR under scrutiny	
Tangnu Romai (44 + 6 MW)	M/s International Ltd. in HPSEC for TEC		5.7.2002			DPR under scrutiny	
Paudital Lassa (24 MW)	M/s Shree Jayalakshmi in HPSEC for TEC		6.6.2002			DPR under scrutiny	
Lambadug (25 MW)	M/s Himachal Consortium		13.6.2002	TEC 17.12.04		The company is in process of obtaining other clearance.	
Baragaon (11 MW)	M/s Kanchanjunga Power Pvt. Ltd.		6.6.2002			DPR under scrutiny in HPSEC	
Sorang (60 MW)	M/s Himachal Sorang Power Pvt. Ltd.		23.9.2004			DPR under scrutiny in HPSEC	
Tidong (100 MW)	M/s Nuziveedu Seeds Ltd.		23.9.2004			DPR under scrutiny in HPSEC	
Budhil (70 MW)	M/s Lanco Green Power Pvt. Ltd.		23.9.2004			DPR under scrutiny in HPSEC	

Source : Official records of HP State Electricity Board, Shimla.

APPENDIX V

Per Capita Plan Investment in Himachal Pradesh

Plan	*Per Capita Annual Investment (in Rs.)*
First Plan (1951-56)	4.00
Second Plan (1956-61)	11.00
Third Plan (1961-66)	21.60
Annual Plan (1966-67)	40.00
Annual Plan (1967-68)	40.00
Annual Plan (1968-69)	40.00
Fourth Plan (1969-74)	61.20
Fifth Plan (1974-78)	100.50
Annual Plan (1978-79)	176.50
Annual Plan (1979-80)	176.50
Sixth Plan (1980-85)	287.80
Seventh Plan (1985-90)	544.59
Annual Plan (1990-91)	765.32
Annual Plan (1991-92)	765.32
Eighth Plan (1992-97)	1353.60
Ninth Plan (1997-2002)	2205.03

Source : Compiled from Plan Documents, Government of Himachal Pradesh.

APPENDIX VI

General Information of SJVN Project Affected Households

Particulars	Number			Percentage		
	SC/ST	Gen.	Total	100	100	100
(1)	(2)	(3)	(4)	(5)	(6)	(7)
1. Total Households	39	106	145	100	100	100
2. Religion						
Hindu	39	106	145	100	100	100
3. Broken Families	2	10	12	5	9	8
4. Assets in possession of Household						
Only Land	4	16	20	10.25	15.10	13.79
Only House	3	1	4	7.70	0.94	2.76
Only Shop	0	1	1	0	0.94	0.69
Land and House	31	84	115	79.49	79.25	79.31
Shop, Land and House	1	4	5	2.56	3.77	3.45
5. Demographic Profile						
Total Population	192	468	660			
Male	99	225	324			
Female	93	243	336			

	Family size	4.92	4.42	4.55			
	Sex ratio	939	1086	1037			
	Marital status	95	268	363			
	Married male	46	131	177	48.42	48.88	48.76
	Married female	49	137	186	51.58	51.12	51.24
6.	Educational Status						
	Illiterate	43	67	110	23.12	14.41	16.90
	Primary	42	83	125	23.58	17.85	19.20
	Middle	47	67	114	25.27	14.41	17.51
	Matriculation	27	100	127	14.52	21.50	19.51
	Senior Secondary	14	59	73	7.53	12.69	11.21
	Graduate	8	58	66	4.30	12.47	10.14
	Post Graduate	5	19	24	2.68	4.09	3.69
	Tech. Diploma	0	12	12	0	2.58	1.84
	Total	186	465	651	100	100	100
7.	Literacy				77	86	83
	Male				85	93	91
	Female				68	78	76

Source : Satluj Jal Vidyut Nigam Ltd.: Baseline Demographic Socio-Economic Survey of Rampur Hydro Electric Project (2006).

APPENDIX VII

Workers and Non-workers in Project Affected Households

Particulars	*Number*			*Percentage*		
	SC/ST	*Gen.*	*Total*	*SC/ST*	*Gen.*	*Total*
(1)	*(2)*	*(3)*	*(4)*	*(5)*	*(6)*	*(7)*
Total population	186	465	651	100	100	100
Male	96	223	339	100	100	100
Female	90	242	332	100	100	100
Workers	102	291	393	54.84	62.58	60.37
Male	48	141	189	50.00	63.23	59.25
Female	54	150	204	60.00	61.98	61.45
Non-Workers	84	174	258	45.16	37.42	39.63
Male	48	82	130	50.00	36.77	40.75
Female	36	92	128	40.00	38.02	38.55
Dependency Ratio				0.82	0.60	0.66

Source : Satluj Jal Vidyut Nigam Ltd.: Baseline Demographic Socio-economic Survey of Rampur Hydro Electric Project (2006).

APPENDIX VIII

Occupational Pattern of Workers of Project Affected Households

Particulars	*Number*			*Percentage*		
	SC/ST	*Gen.*	*Total*	*SC/ST*	*Gen.*	*Total*
(1)	*(2)*	*(3)*	*(4)*	*(5)*	*(6)*	*(7)*
Agriculture						
Male	27	64	91	28.12	28.70	28.54
Female	42	102	144	46.67	42.15	43.37
Agriculture Labour						
Male	4	3	7	4.17	1.34	2.19
Female	1	4	5	1.11	1.65	1.52
Non-agri. Labour						
Male	10	3	13	10.42	1.34	4.07
Female	3	1	4	3.33	0.41	1.20
Service						
Male	7	47	54	7.29	21.07	16.93
Female	0	7	7	0.00	2.89	2.10

(*Contd.*)

Appendix VIII (Contd.)

(1)	(2)	(3)	(4)	(5)	(6)	(7)
Business						
Male	—	20	20	—	8.98	6.27
Female	—	2	2	—	0.83	0.60
Rural Artisans						
Male	—	2	2	—	0.90	0.63
Female	—	0	0	—	0.00	0.00
Students						
Male	37	72	109	38.54	32.29	34.18
Female	25	65	90	27.78	26.86	27.10
Households						
Male	0	0	0	—	0	0.00
Female	8	34	42	8.89	14.05	12.65
Infants						
Male	10	10	20	10.42	4.48	6.27
Female	9	22	31	10.00	9.09	9.34
Un-employed						
Male	0	2	2	0	0.90	0.63
Female	0	0	0	0	0.00	0.00

Non-workers						
Male	1	0	1	1.04	0.00	0.31
Female	2	5	7	2.22	2.06	2.10
Total						
Male	96	223	319	100	100	100
Female	90	242	332	100	100	100

Source : Compiled from Satluj Jal Vidyut Nigam Ltd.: Baseline Demographic Socio-Economic Survey of Rampur Hydro Electric Project (2006).

APPENDIX IX

Source of Income, Extent of Income and Families Below Poverty Line among Project Affected Households

Particulars	Number			Percentage		
	SC/ST	Gen.	Total	SC/ST	Gen.	Total
(1)	(2)	(3)	(4)	(5)	(6)	(7)
			Source of Income			
Service	7	47	54	17.94	44.34	37.24
Agriculture	35	95	130	89.74	89.62	89.65
Business	0	17	17	0	16.03	11.72
Wage labour	28	32	60	71.79	30.18	41.38
Other	0	4	4	—	3.77	2.75
No. of households	39	106	145	100	100	100
			Extent of Income			
Service	23285	53658	45488	26.41	40.46	37.70
Agriculture	51372	44597	46420	58.28	33.63	38.48
Business	308	16150	11880	0.35	12.18	9.85
Wage labour	13182	15150	14621	14.96	11.42	12.12
Other	0	3051	2230	0	2.31	1.85

Total income	88144	132606	120648	100	100	100
Families below poverty line						
Total Households	39	106	145	100	100	100
Households Below poverty line	0	0	0	0	0	0

Source : Compiled from Satluj Jal Vidyut Nigam Ltd.: Baseline Demographic Socio-Economic Survey of Rampur Hydro Electric Project (2006).

APPENDIX X

Ranking of Hill States on the Basis of Composite Index of Physical Infrastructural Development

Sl. No.	Name of Hill State	Composite Index	Rank
1.	Uttaranchal	7.26	1
2.	Nagaland	4.98	2
3.	Himachal Pradesh	4.07	3
4.	Jammu & Kashmir	0.86	4
5.	Manipur	-0.72	5
6.	Tripura	-1.88	6
7.	Mizoram	-2.73	7
8.	Meghalaya	-3.76	8
9.	Sikkim	-3.80	9
10.	Arunachal Pradesh	-4.27	10

Source : Tiwari, A.K. (2007), "Upgradation of Physical Infrastructure" in L.R. Sharma (ed.) *Perspectives of Growth-oriented Hill Economy: Himachal Pradesh*, Shipra Publications, New Delhi.

APPENDIX XI

Indicators of Quality of Life in the States of Himalayan Region

Indicators	*J&K*	*H.P.*	*Sikkim*	*Nagaland*	*Manipur*	*Meghalaya*	*Tripura*	*Arunachal Pradesh*	*Mizoram*	*All-India Average*
(1)	*(2)*	*(3)*	*(4)*	*(5)*	*(6)*	*(7)*	*(8)*	*(9)*	*(10)*	*(11)*
A. Income, Consumption and Equity-related Indicators										
1. Per capita gross state domestic product at 1993-94 (Annual average for 1997-98 to 1999-2000 period)	8454 (19)	11501 (9)	10582 (10)	10182 (12)	8295 (20)	8972 (17)	8109 (23)	9759 (14)	—	10165
2. Annual compound growth rate of per capita gross state domestic product at 1993-94 prices (% per annum in the 1990-91 to 2000-01 period)	2.21	4.45	4.8	0.8	3.91	2.19	5.6	2.18	-	3.83
3. Monthly per capita consumer expenditure in rural sector in Rs. (2000-01)	680	701	620	620	620	620	620	620	620	495
4. % of population below poverty line										
(a) 1973-74	40.83	26.39	50.86	50.51	49.96	50.2	51	51.93	50.32	54.88
(b) 1999-2000	3.83	7.63	32.67	32.67	28.54	33.87	34.44	33.47	19.47	26.1

(*Contd.*)

APPENDIX XI (Contd.)

(1)	(2)	(3)	(4)	(5)	(6)	(7)	(8)	(9)	(10)	(11)
5. % of rural households getting enough food everyday throughout the year (2000-01)	100	99.9	96.3	96.3	96.3	96.3	96.3	96.3	96.3	97.5
B. Literacy and Education-related Indices										
1. Literacy rate (%) age 7+										
(a) Male 1961	19.8	37.6	22.4	27.2	53.5	N.A.	35.3	53.4	N.A.	40.4
2001	65.8	86	76.7	71.8	77.9	66.1	81.5	64.1	90.7	75.9
(b) Female 1961	5.1	11.2	4.9	13	18.9	N.A.	12.4	24.1	N.A.	15.4
2001	41.8	68.1	61.5	61.9	59.7	60.4	65.4	44.2	86.1	54.2
2. Gross enrolment ratios (2000-01) (6-14 years)										
(a) Boys	93.04	98.3	112.97	94.94	95.72	97.29	99.16	107.05	101.35	90.26
(b) Girls	72.23	84.97	113.57	89.15	83.74	93.75	84.57	93.34	91.27	72.36
C. Health-related Factors and Demographic Features										
1. Infant mortality rate (per 1000 of live births)										
(a) Male 1961	78	101	105	76	31	81	106	141	73	122
2000	59	57	50	—	21	65	31	41	18	67
(b) Female 1961	78	89	87	58	33	76	116	111	65	108
2000	46	45	44	—	24	67	39	39	17	69

2. Total fertility rate (1998-99)	2.71	2.14	2.75	3.77	3.04	4.57	—	2.52	2.89	2.85
3. %age of deliveries assisted by health professionals (1998-99)	42.4	40.2	35.1	32.8	53.9	20.6	—	31.9	67.5	42.3
4. % of children age 12-23 months who received all vaccinations (1998-99)	56.7	83.4	47.4	14.1	42.3	14.3	—	20.5	59.6	42
5. %age of under-weight children under three years (1998-99)	34.5	43.6	20.6	24.1	27.5	37.9	—	24.3	27.7	47
D. Factors Affecting Quality of Life of Women and their Empowerment										
1. Sex ratio (women per 1000 men)										
(a) 1961	878	938	904	933	1015	937	932	894	1009	941
(b) 2001	900	970	875	909	978	975	950	901	938	938
2. Percentage of ever married women with any anemia (1998-99)	58.7	40.5	61.1	38.4	28.9	63.3	—	62.5	48	51.8
3. Percentage of women having married by age 18 years (1998-99)	47.5	38.2	35.5	24.4	20.6	34.8	—	39.7	13	64.6

(Contd.)

Appendix XI (Contd.)

(1)	(2)	(3)	(4)	(5)	(6)	(7)	(8)	(9)	(10)	(11)
4. Status of ever-married women's autonomy (1998-99):										
(a) % involved in decision-making on own health care	55.50	80.80	60.20	69.40	43.30	78.90	—	70.00	73.20	51.60
(b) % with access to money	58.10	80.10	78.90	27.90	76.80	81.50	—	78.60	55.00	59.60
5. Percentage of ever-married women of age 15-49 years who were self-employed, or working in family business/ for others (1998-99)	42.50	20.80	22.10	63.90	69.90	47.60	—	59.60	49.90	39.20
6. Percentage of ever-married women who have been beaten or physically mistreated since age 15 years (1998-99)	22.00	5.80	11.40	19.00	19.70	31.10	—	26.40	20.10	21.00
E. Basic Amenities available to Households										
1. Percentage of rural households using electricity for lighting (2000-01)	89.10	95.20	59.80	59.80	59.80	59.80	59.80	59.80	59.80	50.90

2. Perçentage of habitations partially or fully covered under rural water supply (2003)	84.70	98.20	100.00	77.10	99.90	96.20	98.70	92.40	100.00	99.00
3. Average covered area of dwelling unit (in sq.m.) available to rural households (2000-01)	54.10	42.60	52.30	52.30	52.30	52.30	52.30	52.30	52.30	42.50
E. Environmental Factors										
1. Area under forests (in ha.) per capita of rural population										
(a) 1991-92	0.47	0.22	0.70	0.86	0.45	0.65	0.26	6.90	3.50	0.11
(b) 1998-99	0.38	0.20	0.56	0.58	0.35	0.53	0.23	6.10	3.68	0.10
2. Average size of agricultural holdings (ha.)										
(a) 1976-77	1.07	1.62	2.55	7.63	1.11	1.74	1.26	—	—	2.00
(b) 1990-91	0.83	1.20	2.11	6.84	1.23	1.00	0.47	3.71	1.37	1.57

Source : Sharma, L.R. (2004) (ed), "Quality of Life in the Himalayan Region: An Introductory Profile", Indus Publishing Company, pp. 18-24.

Bibliography

Abdul Kalam, A.P.J. (2006), 'PURA in Action' in Verma, S.B., Sankaran, P.N. and Shrivastawa, R.K. (eds.) *Rural-Based Development Strategies*, New Delhi: Deep & Deep Publication, pp. 83-92.

Advisory Board on Energy (1985), "Towards a Perspective on Energy Demand and Supply in India in 2004-05", *Government of India*.

Ahluwalia, Montek S. (2000), 'Economic Performance of States in Post-Reforms Period', *Economic and Political Weekly*, 35(19), pp. 1637-48.

Agrawal, R.N. (2005), "Environmental Concerns and Role of Dams in Development with Special Reference to Bhakra and Beas Projects", Paper presented in *Conference on Development of Hydropower Projects—A Prospective Challenge*, 20-22 April, Shimla.

Alagh, Y.K., Bhalla, G.S. and Kashyap, S.P. (1980), *Structural Analysis of Gujarat, Panjab and Haryana Economies: An Input-Output Study* (New Delhi : Allied Publishers).

Bala Krishna, S. (1979), "Some Demographic Features of Rural India", (ed.) *Rural Development in India*, Hyderabad: National Institute of Rural Development, pp. 51-68.

Bansal, S.P. (2005), "Private Sector Participation in Hydropower Development in Himachal Pradesh", Paper Presented in *Conference on Development of Hydropower Projects—A Prospective Challenge*, 20-22 April, Shimla.

Barnes, D.F. and H.P. Binswanger (1986), 'Impact of Rural Electrification and Infrastructure on Agricultural Changes, 1966-80', *Economic and Political Weekly,* 21 (1)

Barron, W.F. (1992), Rapid Environmental Impact Assessment of Energy Systems, in K.V. Ramani, P. Hills and G. George (eds.) (1992), *Burning Questions: Environmental Limits to Energy Growth in Asian-Pacific Countries During the 1990s,* Asian and Pacific Development Centre, Kuala Lumpur.

Baster, Nancy (1972), "Development Indicators: An Introduction", *The Journal of Development Studies,* Vol. 8, No. 3, pp. 1-20.

Bhandari Laveesh and Kale Sumita (eds) (2007), *Indian States at a Glance, 2006-07 Himachal Pradesh Performance Facts and Figures,* Pearson Power, New Delhi: Pearson Education in South Asia.

Bhatia, M.S. (1999), 'Rural Infrastructure and Growth in Agriculture', *Economic and Political Weekly,* 34(13).

Bhatti, J.P., Singh, Ranveer and Vaidya C.S. (2002), *Impact Assessment of Resettlement Implementation Under Nathpa-Jhakri Hydroelectric Power Project,* A Study Sponsored by Nathpa-Jhakri Power Corporation Ltd., Shimla.

Brahmananda, P.R., and Panchmukhi, V.R. (eds.) (1987), *Development Process of the Indian Economy,* Bombay: Himalayan Publishing House.

Census of India (1986): *Study on Distribution of Infrastructural Facilities in Different Regions and Level and Trends of Urbanisation,* Occasional Paper-1 of 1986, Social Studies Division, Office of the Registrar General, India, Ministry of Home Affairs, New Delhi, pp. 8-49.

Centre for Monitoring Indian Economy (1996), *India's Energy Sector,* CMIE, Mumbai.

Centre for Monitoring Indian Economy (2003), *Energy,* CMIE, Mumbai.

Centre for Monitoring Indian Economy (2004), *Agriculture,* CMIE, Mumbai.

Chand, Ramesh (1996), "Ecological and Economic Impact of Horticultural Development in the Himalayas: Evidence from Himachal Pradesh", *Economic and Political Weekly,* 29 June.

Chatterji, M. (ed.) (1981), *Energy and Environment in the Developing Countries,* New York: Wiley.

Dasgupta, Partha (2001), *Human Well-being and the Natural Environment,* New Delhi: Oxford University Press, p. 13.

Das, Keshab (2006), 'Electricity and Rural Development Linkage', *GIDR Working Paper No. 172,* Gujarat Institute of Development Research, Ahmedabad.

Desai, A.V. (ed.) (1990a), *Patterns of Energy Use in Developing Countries,* New Delhi: Wiley Eastern Ltd.

Desai, A.V. (1990b), *Energy Economics,* International Development Research Centre and United Nations University, New Delhi: Wiley Eastern Ltd.

Dev, Divakar (2001), 'Electrifying Rural India', *Kurukshetra,* 49(5).

Dreze, Jean and Sen, Amartya (eds.) (1996), *Indian Development: Selected Regional Perspectives,* Oxford and Delhi: Oxford University Press.

Dreze, Jean and Sen Amartya (2002), *India: Development and Participation,* New Delhi: Oxford University Press.

Dutt, G. and Ravindranath, N.H. (1992), *Energy End Use: An Environmentally Sound Development Path,* Asian Development Bank, Manila.

Ghosh, P.K. (ed.) (1984), *Energy Policy and Third World Development,* International Resource Books 4, Greenwood Press, Wesport, Connecticut.

Goodman, Raymond (1980), "Managing the Demand for Energy in the Developing World", *Finance and Development,* Vol. 17, No. 4, December 1980, pp. 9-10.

Government of India, *Economic Survey, 2005-06,* Ministry of Finance, New Delhi.

———, (1979), *Report of the Working Group on Energy Policy,* Planning Commision, New Delhi.

Government of Himachal Pradesh, *Draft Annual Plan (2003-04),* Planning Department, Shimla.

———, *Statistical Abstract* (Various Issues), Directorate of Economics and Statistics, Shimla.

———, *Economic Survey* (Various Issues), Directorate of Economics and Statistics, Shimla.

Goldemberg, J. (1990), Solving the Energy Problems in Developing Countries, *The Energy Journal*, Vol. 11, No. 1, pp. 19-24.

Henderson, P.D. (1975), *India: The Energy Sector*, Oxford University Press, New Delhi.

Kothari, C.R. (1991), *Research Methodology (Methods and Techniques)*, New Delhi: Wiley Eastern Limited.

Leontief, W. (1970), Environnmental Repercussions and the Economic Structure: An Input-Output Approach. In W. Leontief (1986) *Input-Output Economics*, 2nd Edition, New York: Oxford University Press.

Mahajan, V.S. (ed.) (1983), *Energy Development in India*, New Delhi: Deep and Deep Publications.

Mohan, Krishna (2005), Addressing Regional Backwardness (An Analysis of Area Development Programmes in India) New Delhi : Manak Publications.

Munasinghe, M. (1988), Energy Economics in Developing Countries: Analytical Framework and Problems of Application, *Energy Policy*, Vol. 9, No. 1, pp. 1-17.

Panchuri, Rajendra K. (ed.) (1980), *Energy Policy for India*, Delhi: The Macmillan Co. of India Ltd.

Pachauri, R.K. and Srivastava, L. (1988), Integrated Energy Planning in India: A Modeling Approach, *Energy Journal*, Vol. 9, No. 4, pp. 35-48.

Pandey, Shridhar (1998), Modern Economic Development and Socio-Cultural Crisis in India, Delhi: Moti Lal Banarasi Das.

Parikh, Jyoti K. (1980), *Energy Systems and Development*, Delhi: Oxford University Press.

Parmar, H.S. (1992), *Tribal Development in Himachal Pradesh*, New Delhi: Mittal Publications.

Paul, S. (1981), Decentralised Energy, *Commerce*, Annual Number 1980.

Pendse, D.R. (1980), "Energy Crisis and its Impact on Energy Consumers in Third Word," *Economic and Political Weekly* Vol. XV, Nos. 3 and 4, January 19 and 26.

Phadke and Sudhir Chella Rajan (2003), "Electricity Reforms in India," *Economic and Political Weekly*, Vol. XXXVIII, No. 29, July 19-25, pp. 3061-72.

Raikhy, P.S. and Parminder Singh (1990), *Energy Consumption in India*, New Delhi: Deep and Deep Publications.

Reliance (2002), *Reliance Review of Energy Markets*, Mumbai: Reliance Industries Limited.

Sen, Amartya, "Capital and Well-being" in Nussbaum, Martha and Smartya Sen (eds) (2002), *The Quality of Life*, New Delhi: Oxford University Press.

Sharma, L.R. (1987), *The Economy of Himachal Pradesh*, Delhi: Mittal Publications,

Sharma, L.R. (ed.) (2002), *Quality of Life in the Himalayan Region*, New Delhi: Indus Publishing Company.

The Tribune (Chandigarh Edition), March 9, 2007.

The Times of India (Chandigarh/Delhi), February 2, 2009.

Talwar, S.P., "Energy Potential", *The Hindustan Times*, April 16, 1981.

TERI (2003), 'Enhancing Electricity Access in Rural Areas through Distributed Generation Based on Renewable Energy', Policy Discussion Forum Base Paper, at http://static.teriin.org/seminar/electricity.pdf

Tiwari, Ashok Kumar (1991), "Infrastructural Development in Himachal Pradesh", *M.Phil. Dissertation*, Department of Economics, Shimla: Himachal Pradesh University).

Tiwari, A.K. (2000), *Infrastructure and Economic Development in Himachal Pradesh*, Indus Publishing Company, New Delhi.

Tiwari, A.K. and Parmar, H.S. (2004), "An Overview of Economic Policy Changes and Economic Development in India", *Nava Arthiki*, Vol. 12, Nos. 1-2.

Tiwari, A.K. Saleem, Mohd. and Garg, Minakshi (2005), " Inter-Regional Disparities in Infrastructural Facilities: A Case Study of Punjab", *Panjab University Research Journal (Social Sciences)*, Vol. 13, Number 3, 2005, pp. 44-57.

Tiwari, A.K. (2005), "Feel Good and Bad Factors in Infrastructure Sector of Himalaya Hill's Economy of India", in S. Murty (ed.), *Feel Good & Bad Factors in Different Sectors*, Jaipur : RBSA Publishers.

Tiwari, A.K. (2006), Exploitation of Hydropower Potential and Economic Benefits to Himachal Pradesh: A Preliminary Observation, *Nava Arthiki*, Vol. 15, No. 2, July-December 2006.

Tiwari, A.K. (2007), *Hydroelectric Power Generation and Economic Development (A Case Study of Nathpa Jhakri Hydroelectric Project of Himachal Pradesh)*, Institute of Integrated Himalayan Studies (UGC Centre of Excellence), Himachal Pradesh University, Shimla.

Tiwari, A.K. (2007), *Spatial Dimensions of Socio-Economic Development in Himachal Pradesh*, Kanishka Publications, New Delhi.

Tiwari, A.K. (2007), "Upgradation of physical Infrastructure for Accelerated Growth of the Hill Economy: Some Suggestions", in Prof. L.R. Sharma (Ed.), *Perspective on a Growth-oriented Hill Economy,* Himachal Pradesh, Shipra Publications, New Delhi.

Tiwari, A.K. (2007), Demography and Literacy in Himachal Pradesh, Jansaksharta, *Journal of State Resource Centre for Adult Education*, Indore, March, 2007.

Tiwari, A.K. and Zinta, R.L. (2007), "Hydropower Generation and Economic Development: A Study of NJPC Project in Himachal Pradesh", *Research Journal Social Sciences,* Vol. 15, No. 2, pp. 136-48

Tiwari, A.K. and Zinta, R.L. (2008), *Socio-Economic Infrastructural Facilities for Sustainable Rural Development: A Factorial Analysis on Himachal Pradesh,* Institute of Integrated Himalayan Studies (UGC Centre of Excellence), Himachal Pradesh University, Shimla.

Tiwari, A.K. (2008), "Economic Infrastructure and Agricultural Development in Himachal Pradesh: A District-level Analysis" *Social Change,* Vol. 38, No. 2, pp. 245-62.

UNDP, *Human Development Report, 2003,* p. 28.

Vaidya, C.S. and Ranveer Singh (2006), Baseline Demographic Socio-Economic Survey of Rampur Hydroelectric Project, A Study Sponsored by Satluj Jal Vidyut Nigam Ltd., Shimla.

World Bank (1979), *India: Economic Issues in the Power Sector,* Washington, D.C.

World Bank (1983b), *The Energy Transition in Developing Countries,* Washington, D.C.

Zinta, R.L. and Tiwari A.K. (2008), *Psychological Analysis on Displaced and Likely to be Displaced Families of Hydroelectric Power Project of SJVN Limited,* Institute of Integrated Himalayan Studies (UGC Centre of Excellence), Himachal Pradesh University, Shimla.

Index